GREEN URBANISM IN ASIA

The Emerging Green Tigers

GREEN URBANISM IN ASIA

The Emerging Green Tigers

Peter Newman • Anne Matan

Curtin University, Australia

World Scientific

NEW JERSEY • LONDON • SINGAPORE • BEIJING • SHANGHAI • HONG KONG • TAIPEI • CHENNAI

Published by

World Scientific Publishing Co. Pte. Ltd.
5 Toh Tuck Link, Singapore 596224
USA office: 27 Warren Street, Suite 401-402, Hackensack, NJ 07601
UK office: 57 Shelton Street, Covent Garden, London WC2H 9HE

Library of Congress Cataloging-in-Publication Data
Newman, Peter, 1945–   .
Green urbanism in Asia : the emerging green tigers / by Peter Newman & Anne Matan (Curtin University, Australia).
pages cm
Includes bibliographical references and index.
ISBN 978-9814425476 (alk. paper)
1. Urban ecology (Sociology)--Asia. 2. City planning--Environmental aspects--Asia. 3. Green movement--Asia. I. Matan, Anne. II. Title.
HT243.A77N49 2013
307.1'216095--dc23

2012046043

British Library Cataloguing-in-Publication Data
A catalogue record for this book is available from the British Library.

In-house Editor: Lee Xin Ying

Typeset by Stallion Press
Email: enquiries@stallionpress.com

Printed in Singapore by Mainland Press Pte Ltd.

Contents

Preface and Acknowledgments

Green urbanism is a recognition that cities make the opportunities that change the world and we need green innovations to save the world. The green and the urban have not always sat easily together. Today, cities must be green for the sake of people, the environment and economies. This is the biggest agenda of our day and we wanted to see the extent to which Asian cities, the economic growth engine of the world, are also taking on green urbanism.

This book was inspired by the work of Professor Tim Beatley from the University of Virginia Charlottesville. His books *'Green Urbanism in Europe'* and the co-written *'Green Urbanism Down Under'* started the tradition into which this book fits.

Our book was constructed with a lot of help from the students completing their Masters in Urban Design at the National University of Singapore. They were the 'forward scouts' in checking out the green urbanism territory in Asia. Their personal insight into their home cities has permeated every page. We hope we have done justice to their aspirations for a greener urban future.

We would sincerely like to thank: Erlangaa Baskara, Imogen Udeshika Jayewardene, Tan Jia Hung, Lim Shu Ying, Yan Liang Liang, Rodeo Calliban, Marianne Pingkan, Raymond Tirtawidjaja, Laraine Macapagal, Ratna Octavia, Ujjwal Parekh, Rahim Hamid, Alia Nur, Arjun Rosha, Chen Hong, Chen Quiong, Tan Yan Liang, Wang Kai Brien, Arizzky Magetsari, Dhanen Mohanadas, Desiree Lee, Pavithra Chandramohan, Han Je, Nicholas Lourdes, Hemalatha Chandramohan, Edo Adrianus Kartono, Bij Borja, Uraiwan Thaveeprungsiporn, Wutt Yee Khin, Amit Gupta, Carmichael Castillo, Emerson Ramos, Jennifer Fiona, Mehnaz Ahmed, Andre Yusak, Gaidhani Saurabh Laxmikant, Susanto Soenjoyo, Harvey Lukman, Irwin Konstantin Ponco, Rocky Yulianto

Saputra, Rulyan Ali Parinduri, Fahry Adhitya, Tan Jia Yu Fiona, Wong Xinni (Faith), Ankit Kumar Kamboj, Xia Wannuo and Rakshada Rode.

Peter Newman and Anne Matan
Curtin University Sustainability Policy (CUSP) Institute

CHAPTER ONE

Introduction

In February 2011 a new book was launched in Singapore called *'How Asia can Shape the World: From the Era of Plenty to the Era of Scarcities'*. The author, Jorgen Oerstrom Moeller, was a former diplomat from Denmark who is a close observer of Asia and who can see that there is a major tide change in global affairs. He believes that the era of scarcity has begun as the limits of the earth are consumed by modern industrial economies based on the American model. The next century, he suggests, will be the time for Asia to demonstrate to the world how to create a different kind of global economy that is more sustainable and is based on "shifting of productivity from the individual to groups, ecological productivity instead of economic productivity, and a reversal to traditional Asian values — less materialistic than Western values".

Our book is written to investigate the first signs of this changing economy. It focuses on cities because that is where the growth and dynamism of change is happening globally and especially in Asia. Here is an opportunity to demonstrate the new paradigm and we went looking to see if we could find the first signs of this as they are unfolding.

This book is a collection of stories as the trends and data are not yet available to really assess the impact of the changes we are observing. The idea of urban stories changing the world is not new. Jane Jacobs developed a view that innovation occurs through urban economies copying each other with each one building better versions (Jacobs, 1984). Stories, especially stories of hope, are powerful as they fire the imagination and give a hint that a better future may be possible. This has been the basis of several other books that Peter Newman has written and been associated with (Beatley and Newman, 2009; Newman and Kenworthy, 1999; Newman *et al.*, 2009).

The ideas that will shape our new urban economies are beginning to emerge as cities across the world begin to grapple with the overlapping issues of global climate change, peak oil, global biodiversity decline, loss of cultures, and the many challenges for housing and infrastructure faced by cities flooded with new migrants from their rural hinterlands. We do not spend time detailing these problems; the literature is full of that. Instead, we want to show that the problems are being understood and the first steps towards a resolution are there, at least in demonstration form.

The hope that we have generated in this book is based on the ideas of innovation waves, that human societies and especially cities have adapted to over the past few hundred years. This notion is expressed in Figure 1.1 showing how economic cycles have followed these waves. It

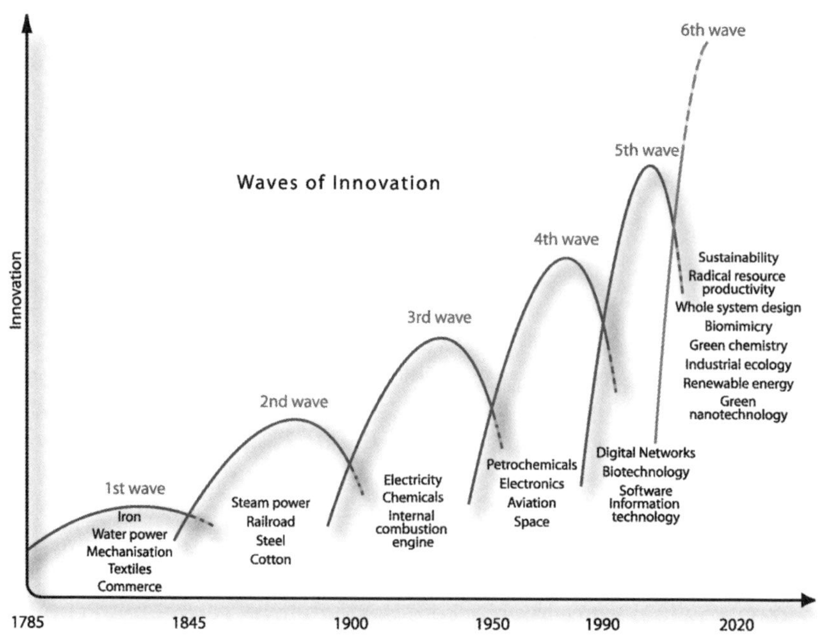

Figure 1.1 Waves of Innovation.
Source: Hargroves and Smith, 2005.

shows how the waves of innovation take off after periods of economic decline and lead to the next form of economy and with it the next form of cities. This approach has been outlined in detail in Newman *et al.* (2009).

The next phase of economic development as shown in Figure 1.1 is a combination of digital and sustainability technologies. But in no period of history has there been a simple change of technology that has renewed and reformed the economy, it has always involved changes in the way business has been done and changes in the way cities have been formed, designed and managed (see Box 1.1).

Box 1.1 Economic Waves, Urban Infrastructure and City Form

The First Wave cities were traditional Walking Cities with some use of horses and carriages while new industries began to develop along rivers and canals using water power.

Then the Second Wave cities spread out along the railways of the steel and steam era. European cities have retained much of this corridor type of city and the remains of the Transit City are still apparent in US and Australian cities with their pearls along a string.

The Third Wave of electricity and the internal combustion engine saw electric tramways built as the basis of most cities, especially in the growing cities of America such as Los Angeles which had the world's most extensive tramway system. These cities followed linear development patterns along the tramways. At the same time the first cars and buses were appearing though they did not begin to dominate city form.

Then the Fourth Wave was dominated by cheap oil, which enabled cities to spread and sprawl for 50 miles or so in every direction. Thus were created the Automobile City forms that confront us today with the biggest challenge of responding to fossil fuel reductions, especially petroleum.

The Fifth Wave of IT and digital technologies has not shaped cities much other than to replace the old industrial manufacturing parts of central and inner areas with knowledge jobs, thus helping to minimize some of the sprawl and start the renewal of these older industrial sites from the previous Waves. However, the Fifth Wave still had cheap oil-enabling cars to dominate cities.

(Continued)

(*Continued*)

The Sixth Wave coincides with the end of cheap oil and the beginning of radical resource productivity of a kind that can deliver 50–80% less fossil fuels by 2050. It also includes a start of a new series of sustainability technologies related to renewables and distributed, small-scale water, energy and waste systems (building on clever control systems now perfected from the Fifth Wave), all of which are more local and require far less fuel to distribute.

This all means the city can become much more polycentric. The transport systems that support such energy and such polycentricity appear to be new electric transit systems for fast cross-city movement and a series of small-scale electric and hybrid vehicles for small local trips as well as walking and cycling which have survived all the city form changes. The polycentric centers and the remaining suburban buildings all need to be renewed as solar-based with all the eco-technologies of the Sixth Wave.

Source: Derived from Newman and Kenworthy, 1999.

This book will try to set out first of all what are the characteristics of the new city in this Sixth Wave. There are many words that are being used to describe this new kind of city — 'sustainable city', 'resilient city' and 'green urbanism' because it is simpler and more easily understood. Thus the book will begin in Chapter 2 to describe what green urbanism is, based on our experience in Western industrial cities. Based on the seven characteristics we have used, it will then seek to show the emerging trends and stories we have found across Asia.

The question underlying this investigation is whether Asian cities will take on or even take over this green urbanism agenda. It is estimated that by 2030, the urban population will reach five billion, or 60%, of the world's population. This is largely contributed to by the urbanization occurring in the Asian region. Asia's cities are predicted to be home to more than 60% of the world's urban dwellers by 2050. This is a threat if Asian cities develop according to the Western Fourth Wave model, but if it uses the growth opportunity to build cities as models of the Fifth to Sixth Wave — the green urbanism era — then they will show the rest of the world how it can be done. The Organization for Economic Co-operation and Development (OECD) and the

United Nations Environment Programme (UNEP) have been saying for the past decade that the next economic agenda is the green economy or green growth (OCED 2011a, 2011b) or green urbanism, as the Asian Development Bank calls it (Asian Development Bank, 2012). This book will look at the urban expression of this new agenda as it merges the Fifth and Sixth Waves of Innovation. Will green urbanism in Asia help shape the world in the age of scarcity?

CHAPTER TWO

What is Green Urbanism?

2.1. Green Urbanism

Green urbanism is often used as a way to describe settlements that are smart, secure and sustainable. They are smart in that they are able to adapt to the new technologies of the 21st century, secure in that they have built-in systems that enable them to respond to extreme events as well as being built-to-last, and sustainable in that they are part of the solution to the big questions of sustainability, such as climate change, peak oil, and biodiversity. Green urbanism has been applied by Beatley (2000) to European exemplar cities and to innovations across Australian cities by Beatley and Newman (2009). But the ideas have not been applied to Asian cities, yet.

This book will use the ideas of Newman *et al.* (2009) on what constitutes the main characteristics of resilience in cities as the overlap with the ideas behind green urbanism are very similar. Seven characteristics are outlined to help define these features and they are illustrated from around the world before applying them to Asian cities.

2.2. Innovations in Green Urbanism

Globally, there are seven features of green urbanism that are emerging. These are described as seven archetypal cities:

- The Renewable Energy City
- The Bioregional Carbon Neutral City
- The Distributed City
- The Biophilic City

- The Eco-efficient City
- The Place-based City
- The Sustainable Transport City.

These city types are obviously overlapping in their approaches and outcomes, but each provides a perspective on how attempts to improve the green urbanism of a city can be achieved. While no one city has shown innovation in all seven areas, some are quite advanced in one or two. The challenge for urban professionals is to apply all of these approaches together, to generate a sense of purpose through a combination of new technology, city design and community-based innovation.

2.2.1. *The Renewable Energy City*

There are now a number of urban areas that are partly powered by renewable energy techniques and technologies, from the regional to the building level. Renewable energy enables a city to reduce its ecological footprint, and if biological fuels are used, it can be part of a city's enhanced ecological functions.

Renewable energy production can and should occur within cities, integrated into their land use and built form, and comprising a significant and important element of the urban economy. Cities are not simply consumers of energy, but catalysts for more sustainable energy paths, and can increasingly become a part of the earth's solar cycle.

While some solar city projects are underway, including Treasure Island in San Francisco, there are presently no major cities in the world that are powered entirely by renewable energy. Movement towards a renewable-energy future will require much greater commitment from cities themselves at all levels, including the local and the metropolitan.

Masdar City in the United Arab Emirates is an important first example of a city built from scratch with 100% renewable energy and zero car use (in theory). It is being built with a 60 MW solar photovoltaic plant to power all construction, and eventually a 130 MW solar photovoltaic plant for ongoing power as well as a 20 MW wind farm and geothermal heat pumps for cooling buildings. Electric automatic pod cars on an elevated structure or underground will be the basis of the transport. Masdar has begun to be built (Revkin, 2008).

North Port Quay in Western Australia is designed to be home to 10,000 households and is designed to be 100% renewable through solar photovoltaics, small wind turbines called wind pods and a nearby wave power system. The development will be dense and walkable, with an all-electric transport system featuring electric public and private transport, all linked to renewable power through battery storage in vehicles. The concept has had a mixed response and will be several years in its planning phase because such developments are not easy to accept into town planning schemes designed around fossil fuel-based development.

Urban planning is necessary to create the infrastructure needed to support solar and wind power at the scale necessary to help power a city. While finding locations for large wind farms near urban areas has been controversial (such as the wind farm proposal that was defeated off the coast of Cape Cod, Massachusetts), there are significant opportunities to harness solar and wind power. Studies are also now showing that wind, like photovoltaic solar power, can be integrated into cities and their buildings.

Hydropower has been used in cities such as Vancouver and Christchurch, New Zealand, for decades. Hydropower is growing slowly due to the impact of large dams, but geothermal power appears to be offering a similar level of base load renewable power.

New model cities that are 100% renewable are needed, but retrofitting existing cities is just as important. Cape Town is moving to become 10% renewable and Adelaide has gone from 0% to 20% renewable energy in ten years by building four large wind farms. In Europe, Freiburg and Hannover have become demonstrations on how to bring renewable energy into city planning (City of Hannover, 1998; Scheurer and Newman, 2008).

The City of Vaxja in Sweden has developed a locally based renewable energy strategy that takes full advantage of its working landscapes, in this case the abundant forests near the city. Vaxja's main power plant, formerly fueled by oil, now depends on biomass almost entirely from wood chips, most of which are a by-product of commercial logging in the region. The wood, more specifically, comes from the branches, barks and tops of trees, and is derived from within a 100 km radius of the power plant. This combined heat and power plant (Sandvik II) provides the entire town's heating needs and much of its electricity needs. Its conversion to using biomass as a fuel has been a key element in the city's aspiration to become an oil-free city. Clearly, each city can develop its own mix of local renewable sources, but Vaxja has demonstrated that it can transition from an oil-based power system to a completely renewable system without losing its economic edge. Indeed, cities that develop such resilience early are likely to have an edge as oil resources decline.

The metropolitan landscape can be viewed as the palette for a creative mix of solar design and renewable energy projects; each city and region will have its own special opportunities and resources and in doing so will help create more resilient cities.

One of the most important potential biofuel sources of the future is blue-green algae that can be grown intensively on rooftops. Blue-green algae photosynthesize, so all that they require is sunlight, water and nutrients. The output from blue-green algae is ten times faster than most other biomass sources, so it can be continuously cropped and fed into a process for producing biofuels or small-scale electricity. Most importantly, city buildings can all use their roofs to tap solar energy for local purposes without the distribution or transport losses so apparent in most cities today. According to one advocate of this approach, "every roof should be photosynthetic", meaning a green roof for biodiversity purposes, water collection, photovoltaic collectors or biofuel algal collectors. This can become a solar ordinance set by town planners as part of local government policy.

Few cities have done much to take stock of their photosynthetic energy potential. Municipal comprehensive plans typically document and describe a host of natural and economic resources found within the boundaries of a city — from mineral sites, through historic buildings, to biodiversity — but estimating incoming renewable energy (sun, wind, wave, biomass or geothermal) is usually not included. In advancing the renewable energy agenda in Barcelona, the city took the interesting step of calculating incoming solar gain. As a former sustainable city counselor noted, this amounted to "ten times more than the energy the city consumes or 28 times more than the electricity the city is consuming" (Puig, 2005). The issue is how to tap into this across the city.

The shift in direction to a renewable city can occur through many actions: demonstration solar or low-energy homes created to show architects, developers, and citizens that green can be appealing; procurement actions that source regionally produced wind and other renewable energy to power municipal transit, lights and buildings; and green building standards and requirements for all new public as well as private buildings.

Along with incentives (financial and otherwise), solar cities recognize the need to set minimum regulatory standards. Barcelona has a solar ordinance, which requires new buildings, and substantial retrofits of existing buildings, to obtain a minimum of 60% of hot water needs from solar. This has already led to a significant growth of solar thermal installations in that city.

Transport can also be a major part of the renewable energy challenge. The more public transport moves to electric power, the more it can be part of a renewable city. Calgary Transit's creative initiative called 'Ride the Wind' provides all the power needed for its light rail system from wind turbines in the south of Alberta, Canada. Private transport can now also be part of this transition through a combination of electric vehicles and new battery storage technology, collectively called 'Renewable Transport' (see Went *et al.*, 2008). Not only can electric vehicles use renewable electricity to power their propulsion, they can also be plugged in during the day and add to the grid through their batteries, because their power systems store four times their consumption. Thus, they can provide a critical role in enabling renewable energy to build up a much higher proportion of the urban energy grid. There is also a growing belief that natural gas, which can be created from carbon dioxide (CO_2) and sunlight, will be the renewable freight and industry fuel of the future.

Renewable power enables cities to create healthy and liveable environments while minimizing the use and impact of fossil fuels. But, by itself, this will not be enough to ensure resilient, green urban development.

2.2.2. *The Bioregional Carbon Neutral City*

Carbon neutral has become something of a rallying cry for those who want to do something about climate change. It can become the goal for all urban development, just as it has become for some local governments, businesses and households. This will require a three-step process:

- Reducing energy use wherever possible — especially in the building and transportation sectors;

- Adding as much renewable energy as possible, while being careful that the production of the renewable energy is not contributing significantly to greenhouse gases; and
- Offsetting any CO_2 emitted through purchasing carbon credits, particularly through tree-planting.

The City of Sydney and the City of Fremantle in Australia are now officially 'carbon neutral', as is the South Fremantle Senior High School which in 2012 became the first Australian school to officially be accredited as carbon neutral (City of Fremantle, 2012; City of Sydney, 2012; South Fremantle Senior High School, 2012). These organizations have gone through a carbon auditing process that enabled them to identify the areas of carbon savings and add in renewable energy options to replace fossil power and fuel, then they have found carbon offsetting projects to neutralize the remaining carbon — mostly in their bioregion. The most important focus for this part of green urbanism is to see how the move towards carbon neutrality through offsetting can be put to use in the bioregions of cities to advance the regeneration of landscapes and the reversal of declines in biodiversity. Thus, this section emphasizes bioregional partnerships in carbon neutral projects, especially in its application to Asian cities.

In 2007, the head of News Corporation, one of the biggest media empires in the world, announced that the company would be going carbon neutral. This has led to some remarkable innovations within the company as it confronted the totally new territory of becoming a global leader in energy efficiency, renewable energy and carbon offsets (see www.newscorporation.com). Many businesses, universities, local governments and households are now committed to minimizing their carbon footprint and even becoming carbon neutral. But can it become a feature of complete cities? There are those who suggest it is essential if the world is to move to 'post-carbon cities' (Lerch, 2007).

Several initiatives focus on helping cities to reach these goals, including ICLEI — Local Governments for Sustainability's Cities for Climate Change program, Architecture 2030, the Clinton Foundation's C-40 Climate Change Initiative and UN-HABITAT's Cities for Climate Change Initiative (CCCI). And as mentioned in the above section, many municipalities have started to offer incentives and/or require that new buildings meet certain green building standards. Minimizing carbon at the building level has momentum, as it is easier to integrate the technology into new buildings, and the benefits have been proven — not just in energy savings, but in increased productivity and fewer sick days in green office buildings (Williams, 2012). Zero energy buildings and homes go well beyond what is required by any green building rating system. These have been built in the Netherlands, Denmark and Germany for at least ten years, and there are now increasingly positive examples in every region of the world (Williams, 2012).

In Sydney, the State of New South Wales, through its Building and Sustainability Index (BASIX) program, has mandated that new homes must now be designed to produce 40% fewer greenhouse gas emissions, compared with an existing house (after initially requiring 20% and finding it relatively easy to achieve), as well as 40% less water use. The program aims at reducing CO_2 emissions by 8 million tonnes and water use by 287 billion litres in ten years (Farrelly, 2005). This is an important role for urban planning which can help to set up carbon neutral suburbs through the assessment process. The next phase of this project is called PRECINX and seeks to establish the statutory planning governance for carbon neutral neighborhoods and subdivisions. A model to determine this has been developed by Kinesis and is being used to show how urban developments can approach carbon neutrality (Beattie *et al.*, 2012).

The United Kingdom government has decided that all urban developments will be carbon neutral by 2016, phasing in from 2009. The Beddington Zero Energy Development initiative is the first carbon neutral community in the United Kingdom. It has extended the concept to include building materials and, as it is a social housing development, it has shown how to integrate the carbon neutral agenda with other sustainability goals, making it a more resilient demonstration.

Malmö, Sweden, has stated that it has already become a carbon neutral city; Vaxja, Sweden, has declared its intention to become a fossil fuel-free city, and Newcastle in the United Kingdom and Adelaide also aspire to be carbon neutral. Each has taken important steps in the direction of renewable energy. Vancouver's new Winter Olympic Village was built as a model of North American demonstration in carbon neutral urban development.

The link to the green agenda of a city is very direct with respect to the carbon neutral approach of bioregional tree-planting schemes. By committing to be carbon neutral, cities can focus their offsets into bioregional tree-planting, as part of the biodiversity agenda as well as to address climate change.

In all Australian cities, the carbon and greenhouse gas emissions associated with many municipal motor pools are being offset through innovative tree-planting initiatives and through organizations such as Green Fleet, which has recently planted its two-millionth tree. Firms, such as airlines, offer carbon neutral services and schools as well as many businesses are committed to being carbon neutral. The carbon offsetting is accredited through a Federal Government scheme called 'Greenhouse Friendly' and provides a strong legal backing to ensure that tree-planting is real, related to the money committed and guaranteed for at least 100 years as required by the Kyoto Convention. The City of Fremantle in Western Australia declared itself carbon neutral in 2010.

Many of the carbon offsetting programmes associated with carbon neutral programmes are going towards biodiversity plantations that are regenerating the ecology around cities. A particular example is the Gondwana Links project, which is regenerating an ecological link, or corridor, over 3,000 km long between the coastal ecosystems of the Karri forest to the inland woodlands by joining up various reserves across the whole south coast of Western Australia. The project is driven by many big firms, as well as ordinary people, using their carbon offsets from energy use to create this biodiversity-based tree-planting (Newman and Jennings, 2008).

Preserving and planting trees helps to sequester carbon that is emitted. Tree cover also helps to cool buildings and homes naturally and can reduce the use of energy for artificial cooling. Tree-planting as part of a carbon neutral program can be built into the normal functions of a city, as in Cairo (Duquennois and Newman, 2009; UN-HABITAT, 2009). Other urban initiatives to provide greater tree coverage through offsets are the Sacramento Municipal Utility District in California, Atlanta's beltway project and Los Angeles' 'One Million Trees' program.

Progress in moving away from fossil fuels also requires serious localizing and local sourcing of building materials. This, in turn, provides new opportunities to build more photosynthetic economies. The value of emphasizing the local is many-fold and the essential benefits are usually clear. The dramatic reduction in the energy consumed as part of making these materials is, of course, the primary benefit. It is also about strengthening local economies and helping them to become more resilient in the face of global economic forces and it is also about re-forming lost connections to place.

At the Beddington Zero Energy Development project in London, more than half of the building materials for the project came from within a 60-km radius, and the wood used in construction, as well as the fuel in the neighborhood's combined heat and power plant, comes from local council forests. A bioregional carbon neutral approach to cities can produce local fibre which will mean an added reduction in fibre miles as well as potential to help re-grow local bioregions.

No city is yet committed to a comprehensive city-wide carbon neutral approach that can link tree-planting to a broader biodiversity cause. If this is done, cities can raise urban and regional reforestation to a new level and contribute to reducing the impact of climate change, simultaneously addressing local and regional green agenda issues.

The carbon neutral city will receive a big boost when a global compact on carbon trading can be achieved, because this will enable the voluntary carbon trading market to become mainstream. By 2012, fifty-five nations will have a carbon pricing regime in place, with Europe as the largest example. Globally, the Clean Development Mechanism is acting as a means to create offsets in the developing world. The use of Kyoto-compliant carbon credit schemes can become a part of any city and its bioregion.

2.2.3. *The Distributed City*

The development of distributed power and water systems aims to achieve a shift from large centralized power and water systems to small-scale and neighborhood-based systems in cities. The distributed use of power and water can enable a city to reduce its ecological footprint, because power and water can be more efficiently provided using the benefits of electronic control systems and community-oriented utility governance.

Most power and water systems for cities over the past 100 years have become bigger and more centralized. While newer forms of power and water systems are increasingly smaller in scale, they are often still fitted into cities as though they were large. The movement that tries to see how these new technologies can be fitted into cities and decentralized across grids is called 'distributed power and distributed water systems' (Droege, 2006).

The distributed water system approach is often called 'water-sensitive urban design'. It includes using the complete water cycle, that is, using rain and local water sources like groundwater to feed into the system and then to recycle 'grey' water locally and 'black' water regionally, thus ensuring that there are significant reductions in water used. This system can enable the green agenda to become central to the infrastructure management of a city; stormwater recycling can involve swales and artificial wetlands that can become important habitats in the city. Grey water can similarly be recycled to irrigate green parks and gardens, and regional black water recycling can be tied into regional ecosystems. All these initiatives require 'smart' control systems to fit them into a city grid and also require new skills among town planners and engineers, who are so far used to water and energy management being a centralized function rather than being a local planning issue (Benedict and McMahon, 2006).

There are now many cities that are able to demonstrate small-scale local water systems that are very effective. The many cities in developing countries that already have distributed water supplies from community boreholes and small-scale sewage treatment can look to a number of cases where these have been made safe and effective without being turned into expensive centralized systems (Ho, 2002). In Malang, East Java, a small-scale community sewage system was fitted into a squatter village to provide sanitation for 500 families (Newman and Jennings, 2008). The City of Sydney is now building on work started in London and Woking by Allan Jones on distributed energy (City of Sydney, 2012). They suggest savings of 70–80% in greenhouse gas are feasible through local trigeneration or cogeneration schemes supplemented by photovoltaics (PV) on rooftops and energy efficiency in buildings (City of Sydney, 2012). Similar work using 21st century design tools has demonstrated 58% savings for greenhouse gases (GHG) and 74% water savings in the Cockburn Coast project in Perth (Beattie *et al.*, 2012).

In large cities, the traditional engineering approach to providing energy has been through large centralized production facilities and extensive distribution systems that transport power over relatively long distances. This is wasteful because of line losses; also, because large base-load power systems cannot be turned on and off easily, there is considerable power shedding when the load does not meet the need. Renewable, low-carbon cities aim at developing more decentralized energy production systems, where production is more on a neighborhood scale, and both line losses and power shedding can be avoided. Whether from a wind turbine, a small biomass combined heat and power plant (as in London's new distributed energy model), or a rooftop photovoltaic system, renewable energy is produced closer to where it is consumed and, indeed, often directly by those who consume it. This distributed generation offers a number of benefits, including energy savings, given the ability to better control power production, lower vulnerability and greater resilience in the face of natural and human-made disasters (including terrorist attacks). Clever integration of these small systems into a grid can be achieved with new technology control systems that balance the whole system in its demand and supply from a range of sources as they rise and fall, linking it to storage, especially vehicle batteries through vehicle-to-grid (V2G) technology. A number of such small-scale energy systems are being developed to make cities more resilient in the future (Sawin and Hughes, 2007).

Distributed power and water provision in cities needs new local or district utilities to manage it and it needs community support as well. In Toronto, a possible model similar to those outlined above in developing cities has been created. Communities began to form 'buying-cooperatives' in which they pooled their buying power to negotiate special reduced prices from local photovoltaic companies that had offered an incentive to buy solar photovoltaic panels. The first cooperative was the Riverdale Initiative for Solar Energy. In this initiative, 75 residents joined together to purchase rooftop photovoltaic systems, resulting in savings of about 15% in their purchase cost. This then spread across the city. The Toronto example suggests the merits of combining bottom-up neighborhood approaches with top-down incentives and encouragement. The support for small-scale distributed production is offered through what are commonly referred to as standard offer contracts (often referred to as 'feed-in tariffs' in Europe); such productions have been extremely successful in Europe, where they are now common. The same can be done with new technologies for water and waste, such as rainwater tanks and grey water recycling.

Another model example is the redevelopment of the Western Harbor in Malmö, Sweden. Here the goal was to achieve distributed power and water systems from local sources. This urban district now has 100% renewable power from rooftop solar panels and an innovative stormwater management system that recycles water into green courtyards and green rooftops. The project involves the local government in the management and demonstrates that a clear plan helps to drive innovations in distributed systems (City of Malmö, 2005).

Distributed infrastructure is beginning to be demonstrated in cities across the globe. Utilities will need to develop models with city planners of how they can carry out local energy and water planning through community-based approaches and local management.

2.2.4. *The Biophilic City*

Biophilic cities are landscaping buildings and using natural systems such as urban creeks to help bring more nature into cities and to help 'air condition' the city. Green roofs and green walls are two most obvious new techniques that have emerged to demonstrate the biophilic city but older techniques such as water-sensitive urban design and daylighting of streams have also become more and more important. Growing energy and providing food and bio-materials locally are also becoming part of urban infrastructure development and building design. The use of biophilic processes in cities reduces their ecological impact by replacing fossil fuels and can bring substantial ecological benefits through an emphasis on natural systems (Beatley, 2011). Timothy Beatley, an American urban planner, promotes the vision of biophilia for a city in his book '*Biophilic Cities*' (2011), defining it as "cities with a society that live much more in harmony with nature". He challenges us to rethink urban infrastructure in an effort to eventually develop a multi-level design that connects humans with nature. He believes that a biophilic city cherishes the natural features that already exist and helps to restore and repair what has been lost or degraded.

Similarly, visions for a more biophilic city have crossed many minds and have been depicted in drawing, animations and the arts. For instance, Luc Schuiten, an eco-visionary architect, has long been interested in the values of sustainability and ecology in the designs of his buildings since the mid-seventies. In his Vegetal City, he demonstrates that a long-lasting and bright future is possible for the planet, through an innovative process.

Reductions in energy requirements for buildings and the cooling of the urban atmosphere have become important in cities not just because of the need to save fossil fuels but because of the urban heat island effect, which has been getting worse as climate extremes become more common (Frumkin, 2008; Preston and Jones, 2006). Biophilic systems covering buildings act as an insulating layer to reduce energy needs in buildings and also to slow down rainfall as it becomes stormwater. This has been a major goal for some cities where pollution is created by the surges in stormwater, as well as flash flooding.

Renewable energy can be tapped from the sun, the wind and geothermal sources using small-scale decentralized technology, as described in the previous section. However, renewable energy can also be grown through biofuels. The transition to growing fuels is drawing on crops and forests that can feed into new ways of fueling buildings and vehicles. Farms and open areas around cities are being developed as the source of renewable energy, especially in the production of biofuels. However, biofuels are also being produced as part of improving the urban environment. This means more intensive greening of less dense parts of cities and their peri-urban regions with intensive food growing, renewable energy crops and forests, but greening the high-density parts of cities as well.

One of the major rationale for biophilic systems being incorporated into high-density parts of cities is the growing belief that a "daily dose of nature" as Timothy Beatley calls it, is a necessary part of how human beings should live (Beatley, 2011). Biophilia as described by Wilson (1984) is the idea that humans co-evolved with nature and need to have natural ecosystems in their daily life. What this actually means is not yet tested and is likely to vary considerably. However the idea is growing that blocks of high-rise buildings without any connection to natural systems may need to change. Biophilic urbanism is emerging as the set of techniques that enables this to occur (Figure 2.1).

As well as natural systems being incorporated into, on top of and around buildings for aesthetic and other landscaping purposes, cities can incorporate food into this biophilic view of the future. Food, in the globalized market place, increasingly travels great distances — apples from New Zealand, grapes from Chile, wine from South Australia, vegetables from China. 'Food miles' are rising everywhere and already food in the United States travels between 2,500 km and 4,000 km from where it is grown to where it is consumed. Any exotic sources of food come at a high-energy cost. The growing, processing and delivering of food in the United States consumes vast amounts of energy on par with the energy required to power homes or fuel cars (Starrs, 2005). Urban agriculture is thus emerging as part of this biophilic approach.

Figure 2.1 Biophilic Urbanism in Singapore — Global Best Practice.
Note: Singapore is probably the leading city in Asia, and even in the world, demonstrating that high density can be associated with intensive natural features.
Source: Peter Newman.

There are now good examples of new neighborhoods and development projects that design, from the beginning, spaces for community gardens that attempt to satisfy a considerable portion of food needs on-site or nearby. Growing food within cities and urban (and suburban) environments can take any number of forms. Community gardens, urban farms, and edible landscaping are all promising urban options (Halweil and Nierenburg, 2007). Prominent and compelling examples of edible urban landscaping have shown that it is possible to trade hardscape environments for fruit trees and edible perennials. In the downtown Vancouver neighborhood of Mole Hill, for instance, a conventional alleyway has been converted to a green and luxurious network of edible plants and raised-bed gardens, in a pedestrianized community space, where the occasional automobile now seems out of place. New urban development can include places (rooftops, sideyards, backyards) where residents can directly grow food. This has been a trend in developed cities, as new urban ecological neighborhoods have included community gardens as a central design element (for example, Viikki, Helsinki; South False Creek, Vancouver; and Troy Gardens, Madison), but this trend is perhaps most famous in Cuban cities over the past few decades, in response to being cut off from oil imports (Murphy, 1999).

Urban agriculture is also widespread in other developing country cities, where it provides food and income for many poor households. Cities need to find creative ways to promote urban farming where feasible, without creating tension with redevelopments for reduced car dependence through increased density. This may mean that a city can use the many vacant lots for commercial and community farms in areas that have been blighted (for example the estimated 70,000 vacant lots in Chicago). However, if these areas are well served with good transit and other infrastructures, such uses should be seen as temporary and, indeed, can be part of the rehabilitation of an area, leading to the development of eco-villages that are car-free and models of solar building, as in Vauban. Many cities have embarked on some form of effort to examine community food security and to promote more sustainable local and regional food production. These can be integrated into ecologically sustainable urban and regional rehabilitation projects (Beatley, 2005) and can utilize the intensive possibilities of urban spaces, as in urban permaculture (see Newman and Jennings, 2008).

In Madison, Wisconsin, Troy Gardens emerged from excess land owned by a state-owned mental hospital. Dubbed the Accidental Eco-village by those involved in its transformation, the land was being sold in 1995 when the community who used it as a garden and park stepped in and formed an association to try and buy the land. Through partnerships with other non-government organizations and the University of Madison's Department of Urban and Regional Planning, the Friends of Troy Gardens was able to create a diversity of uses that enabled the money to be found. Thus, on the site now is a mixed income co-housing project involving 30 housing units, a community garden with 320 allotments, an intensive urban farm using traditional Hmong agricultural techniques for a community-supported agriculture enterprise, and a prairie restoration scheme which is regenerating local biodiversity (Campbell and Salus, 2003).

2.2.5. *The Eco-efficient City*

In an effort to improve eco-efficiency, cities and regions are moving from linear to circular or closed-loop systems, where substantial amounts of their energy and material needs are provided from waste streams. Eco-efficient cities reduce their ecological footprint by reducing wastes and resource requirements.

A more integrated notion of energy and water entails seeing cities as complex metabolic systems (not unlike a human body) with flows and cycles where, ideally, the things that have traditionally been viewed as negative outputs (e.g. solid waste, wastewater) are re-envisioned as productive inputs to satisfy other urban needs, including energy. The sustainability movement has been advocating for some time for this shift away from the current view of cities as linear resource-extracting machines to an integrated urban metabolism approach. This is often described as the eco-efficiency agenda (Girardet, 2000).

The eco-efficiency agenda has been taken up by the United Nations and the World Business Council on Sustainable Development, with a high target for industrialized countries of a ten-fold reduction in consumption of resources by 2040, along with rapid transfers of knowledge and technology to developing countries. While this eco-efficiency agenda is a huge challenge, it is important to remember that throughout the industrial revolution of the past 200 years, human productivity has increased by 20,000%. The next wave of innovation has a lot of potential to create the kind of eco-efficiency gains that are required (Hargrove and Smith, 2006; Hawkens *et al.*, 1999).

The urban eco-efficiency agenda includes the 'cradle to cradle' concept for the design of all new products and includes new systems like industrial ecology, where industries share resources and wastes like an ecosystem (McDonaugh and Braungart, 2002). Good examples exist in Kalundborg, Denmark; and Kwinana, Australia (Newman and Jennings, 2008).

The view of cities as a complex set of metabolic flows might also help to guide cities dealing with situations (especially in the shorter term) where considerable reliance on resources and energy from other regions and parts of the world still occurs. Policies can include sustainable sourcing agreements, region-to-region trade agreements and urban procurement based on green certification systems, among others. As well, every new development should undergo an assessment process that enables its metabolism to be reduced fundamentally. Embracing a metabolic view of cities and metropolitan areas takes global governance in some interesting and potentially very useful directions.

This new paradigm of sustainable urban metabolism (seeing them as complex systems of metabolic flows) requires profound changes in the way cities and metropolitan regions are conceptualized as well as in the ways they are planned and managed. New forms of cooperation and collaboration between municipal agencies and various urban actors and stakeholder groups will be required. Municipal departments will need to formulate and implement integrated resource flow strategies. New organizational and governance structures will be necessary, as well as new planning tools and methods. For example, municipal authorities can map the resource flows of their city and region to see how they can be part of a comprehensive eco-efficiency plan.

One extremely powerful example of how this eco-efficiency view can manifest in a new approach to urban design and building can be seen in the dense urban neighborhood of Hammarby Sjöstad, Stockholm. Here, from the beginning of the planning of this new district, an effort was made to think holistically, to understand the inputs, outputs and resources that would be required and that would result. For instance, about 1,000 flats in Hammarby Sjöstad are equipped with stoves that use biogas extracted from wastewater generated in the community. Biogas also provides fuel for buses that serve the area. Organic waste from the community is returned to the neighborhood in the form of district heating and cooling. There are many other important energy features in the design as well. The neighborhood's close proximity to central Stockholm and the installation (from the beginning) of a high-frequency light rail system have made it truly possible to live without a private automobile (there are also 30 car-sharing vehicles in the neighborhood). While not a perfect example, it represents a new and valuable way of seeing cities, and requires a degree of interdisciplinary and inter-sectoral collaboration in the planning system that is unusual in most cities (Newman *et al.*, 2009).

Eco-efficiency does not have to involve just new technology; it can also be introduced into cities through intensive use of human resources, as in Cairo's famous Zabaleen recycling system (Duquennois and Newman, 2009). There are many other examples of how cities across the third world have integrated waste management into local industries, buildings and food production (Hardoy *et al.*, 2001).

2.2.6. *The Place-based City*

Sustainability is becoming increasingly understood as a way of developing local economies, as a way of building on a local area's unique and special sense of place, and as a way of nurturing a high quality of life and a strong commitment to community. The more place-oriented and locally self-sufficient a city's economy is, the more it will reduce its ecological footprint and the more it will ensure that its valuable ecological features are enhanced. Place-based city concepts will increasingly be the people-oriented motivation for the infrastructure decisions that are made in each of the other city types.

Local economic development has many advantages in the context of sustainable development including the ability of people to travel less as their work becomes local. Finding ways to help facilitate local enterprises becomes a major achievement for cities in moving towards a reduced ecological footprint. Initiatives designed to help small towns in the United States to grow their own jobs have been developed (Sirolli, 1999). An approach for creating local enterprises that build on the passions and resources of the local community and support local businesses in their early vulnerable steps has also been developed. The inaugural Enterprise Facilitation project, which is designed to create local jobs, was pioneered in the small rural town of Esperance, Western Australia, in 1985, but has since spread across three continents. The success of this initiative is reflected in the words of its chairperson (Sirolli, 1999):

> *We are proud to say almost 800 businesses — or 60% of the entrepreneurs — we met are still running successful, sustainable operations and have contributed more than $190 million in revenue to the local economy. . . . We have averaged almost 40 new business start-ups a year consistently in the last 20 years, which is quite a track record given Esperance has a population of just 13,500 people.*

What the pioneers of these initiatives have found, time and time again, is that place really matters. When people belong and have an identity in their town or city, they want to put down their roots and create local enterprises.

Local economic development is a first priority for most cities. As part of this, many cities are placing increasing emphasis on local place identity, as social capital has been found to be one of the best ways to predict wealth in a community (Putnam, 1993). Thus, when communities relate strongly to the local environment, the city's heritage and its unique culture, they develop a strong social capital of networks and trust that forms the basis of a robust urban economy.

This approach to economic development, which emphasizes place-based social capital, has many supporters, but very few relate this to the sustainability agenda in cities. For example, energy expenditures — by municipalities, companies and individuals — represent a significant economic drain, because they often leave the community and region. Producing power from solar, wind or biomass in the locality or region is very much an economic development strategy that can generate local jobs and economic revenue from land (farmland) that might otherwise be economically marginal, in the process recirculating money, with an important economic multiplier effect. Energy efficiency can also be an economic development strategy. For example, research on renewable energy and the creation of related products have developed into a strong part of the economy in Freiburg, Germany.

All the efforts at localizing energy, food, materials and economic development remain dependent on the strength of the local community. The Beddington Zero Energy Development project shows the critical importance of thinking beyond the design of buildings and seeing urban development through a more holistic community-oriented design lens. However impressive the passive solar design and smaller energy demands of this project are (300 mm insulation, an innovative ventilation and heat recovery system, for instance), much of the sustainability gains will come from how residents actually live in these places. Here, residents are being challenged to rethink their consumption and mobility decisions — there is a car-sharing club on-site, for example, a food-buying club, and a community of residents helping each other to think about creatively reducing their ecological impacts and footprints is emerging. This is actually a hallmark of European green projects and an important lesson for projects elsewhere.

A study that examined a range of European urban ecology innovations concluded that when the innovations came from a close and committed community, they became ingrained in people's lifestyles, giving the next generation a real opportunity to gain from them.

However, many architect-designed innovations that were imposed on residents without their involvement tended to fall into neglect or were actively removed (Scheurer, 2003).

Creating a sense of place is about generating pride in the city about all aspects of the economy, the environment and the culture. The Magic Eyes project in Bangkok has illustrated how a solid waste project can be facilitated through sense of place (UN-HABITAT, 2009). In this project awareness of littering in rivers was dramatized by boats with large painted eyes plying the rivers to remind people that they were being watched if they disposed of rubbish in the river.

Sense of place in a city requires paying attention to people and community development in the process of change — a major part of the urban planning agenda for many decades. This localized approach will be critical to creating a resilient city. It creates the necessary innovations as people dialogue through options to reduce their ecological footprint, which in turn creates social capital that is the basis for ongoing community life and economic development (Beatley, 2005; Beatley and Manning, 1997). City dwellers in many countries already increasingly want to know where their food is grown, where their wine comes from and where the materials that make up their furniture come from. This can increasingly move towards every element of the built environment. Thus, as well as a slow movement for local foods, a slow fibre and slow materials movement for local fabric and building purposes can also help create a sense of place and greater resilience.

City economies in the past had their own currencies and it has been argued that national currencies often fail to express the true value of a city and its bioregion (Jacobs, 1984). Transforming urban economies towards a more bioregional focus has been assisted in some places by adopting complementary currencies that provide an alternative to national currencies and by establishing local financial institutions. It has been argued that a complementary local currency not only facilitates change, but also creates a community with a mutual interest in productive exchange among its members in the bioregion (Korten, 1999). In this way, a community affirms its identity and creates a natural preference for its own products. More than 1,000 communities around the world have issued their own local currencies to encourage local commerce. How this has been related to urban planning is set out in Newman and Jennings (2008) using the example of Curitiba.

Most developed cities have created development bonuses similar to Curitiba's that are part of the non-monetary economy of the city. For example, the City of Vancouver requires that 5% of the value of a development be directed into social infrastructure. This is worked out by the developer and council in discussion with the local community, which may want more landscaped streetscapes, more pedestrian areas, a community meeting space, or even an art-house cinema. Social housing is worked out on the basis of receiving a density bonus for more development rights. The more Vancouver exercises these complementary currency requests, the more the development process works to create better public spaces to go with the private spaces developed by the market. Thus, sustainability can be made to mean something at a very local level through the planning system.

All cities have the opportunity through their planning systems to create their own currencies that work in a parallel but complementary way with normal money. These 'sustainability credits' are not owned by the developer or by the city but they are in fact owned by the community, because their values are expressed in the development bonuses granted. Thus, cities can create community banks of sustainability credit through their planning systems. Most cities in developing countries do not have much to invest in their public spaces; hence, the whole city's economy suffers. Curitiba showed how cities could break that mould.

Through the planning system, cities can create their own sustainability currencies for what they most need as determined by their local citizens — they just need to define them as 'development rights'. These new 'sustainable development rights' could be related to biodiversity credits, greenhouse reduction credits, salinity reduction credits, affordable housing credits or anything else that a community can create a 'market' for in their city and its bioregion. Increasingly, these credits can be related to the mainstream economy (just as carbon credits are) in order to fund the infrastructure needed for the resilient city.

Perhaps the best way that cities have been reclaiming a sense of place is through transforming their urban waterways. Special places, even sacred spaces, are usually associated with the land-water interface. Hence, reclaiming waterfronts and beaches, regenerating rivers and small creeks, has often been the way that communities have expressed their desire to be more sensitive to their local place. In San Francisco the regeneration of the waterfront was enabled after the Embarcadero Freeway was removed. The result is a place that is iconic to locals and a boost for the local economy. In Fremantle a local beach was restored as it was a sacred place for Indigenous people and also where the first white settlers arrived. It had become degraded through port activities and was reclaimed by the local government working with a group of local artists who helped to design and rebuild the foreshore areas as a public space. A public monument outlining the history of the area enables the public to not only enjoy the space but to learn the history and heritage of the area.

2.2.7. *The Sustainable Transport City*

Transport is the most fundamental infrastructure for a city, because it creates the primary form of the city (Kostof, 1991; Newman and Kenworthy, 1999). Cities, neighborhoods and regions are increasingly being designed to use energy sparingly by offering walkable, transit-oriented options, more recently supplemented by vehicles powered by renewable energy. Cities with more sustainable transport systems have been able to increase their resilience by reducing their use of fossil fuels, as well as through reduced urban sprawl and reduced dependence on car-based infrastructure.

The agenda for large cities now is to have more sustainable transport options so as to reduce traffic whilst halving the emission of greenhouse gases by 2050 at least, in line with the global agenda set through the Intergovernmental Panel on Climate Change. For many cities, the reduction of car use is not yet on the agenda, apart from seeing it as an ideal to which they aspire. Unfortunately, for most cities, traffic growth has been continuous and appears to be unstoppable; though the last decade has seen car use peak and decline in many western cities (Newman and Kenworthy, 2011). To reduce a city's ecological footprint and enhance its liveability, it is necessary to manage the growth of cars and trucks and their associated fossil fuel consumption.

The variations in private transport fuel use across 84 cities by Kenworthy and Laube (2001) illustrate that there is a very large difference in how cities use cars and petroleum fuels. Singapore is at the lower end on a per capita basis with just 12 GJ per person compared to 103 GJ in Atlanta and 34 GJ in Perth.

A number of studies have shown that these variations have little to do with climate, culture or politics; even income is very poorly correlated with these variations, but they have a lot to do with the physical planning decisions that are made in those cities (Kenworthy *et al.*, 1999; Newman and Kenworthy, 1999). Debate about the relative importance of urban planning parameters is constant, though within the profession there is increasing awareness that sustainable transport will only happen if much greater attention is paid to urban form and density; infrastructure priorities, especially relative commitment to public transport compared to cars; and street planning, especially provision for pedestrians and cyclists as part of sustainable mobility management.

2.2.7.1. *Urban Form and Density Planning*

Very high-density city centers mean that most destinations can be reached with a short walk or they can have highly effective public transport opportunities due to the concentration of people near stations. If densities are generally lower, but higher along corridors, it is still feasible to have a good transit system. If, however, low densities are the dominant feature of a city, then most activity needs to be based around cars, because only cars can enable people to reach their destinations in a reasonable time. Public transport is not competitive when there are not enough people to justify reasonable services. Most low-density cities are now trying to increase their densities in order to reduce their car dependence (Newman and Kenworthy, 2011; Newman *et al.*, 2009).

Density is a major tool available to planners in cities. It is best used where a city has good transit or wants to build transit, as the resulting transit-oriented developments (TODs) can reduce car use per capita among residents by half and save households around 20% of their income, because they have on average one less car (often none) (Cervero, 2008). TODs are an affordable housing strategy as well. In the United States, shifting 60% of new growth to compact patterns would reduce CO_2 emissions by 85 million metric tons annually by 2030 (Ewing *et al.*, 2007). TODs reduce the ecological footprint of cities and undermine the kind of car-based sprawl that eats into the green agenda of cities. Thus, the TODs strategy can enable a city to put in place a clear urban growth boundary and to build a green wall for agriculture, recreation, biodiversity and the other natural systems, creating a far more resilient city.

If cities are dense, as in many developing countries, but do not have adequate public transport and allow too much traffic to develop in their streets, they can easily develop dysfunctional transport systems. However, their density will always enable them to provide viable public transport solutions if they invest in them, whereas low-density cities are always struggling to provide other options. High density means easier non-car-based access, but it can also mean much greater congestion whenever vehicles are used. If the vehicles in these confined spaces are poorly maintained diesel engines, then serious air pollution can result, so cities need to be very serious about managing the source of such emissions (Jain, 2004; Rosencranz and Jackson, 2002; UN-HABITAT, 2009).

2.2.7.2. *Infrastructure Priorities and Transit Planning*

The 'transit-to-traffic' ratio measures how effective public transport is in competing with the car in terms of speed. The best European and Asian cities for transit have the highest ratio of transit to traffic speeds and have achieved this invariably with fast rail systems (Kenworthy, 2008). Rail systems are faster in every city in the study sample by 10–20 km/h over bus systems that rarely average faster than 20–25 km/h. Busways can be quicker than traffic in car-saturated cities, but in lower-density car-dependent cities, it is important to use the extra speed of rail to establish an advantage over cars in traffic. This is one of the key reasons that railways are being built in more than 100 United States cities. In many other cities, modern rail is now seen as the solution for reversing the proliferation of the private car. Rail is also important because it has a density-inducing effect around stations, which can help to provide the focused centers so critical to overcoming car dependence, and they are also electric, which reduces vulnerability to oil as almost no city uses oil for electricity.

Across the world, cities are building modern electric rail systems at vastly increasing rates, because they simultaneously address the challenges of fuel security, decarbonize the economy in the context of addressing climate change, reduce traffic congestion sustainably, and create productive city centers. The trend towards fast electric rail in cities is now being called a 'Mega Trend' (Rubin, 2009). Chinese cities have moved from their road building phase to building fast modern railways across the nation. China is committed to building 120,000 km of new rail by 2020. Investment has risen from 155 billion *yuan* (US$22 billion) per year in 2006 to 1,000 billion *yuan* per year in 2009 (US$143 billion), with around 6 million jobs involved. These projects are part of China's response to the recent global economic downturn (Xin, 2008). Beijing now has the world's biggest metro.

Delhi is building a modern electric metro rail system, which has considerably boosted the city's pride and belief in the future. The 250 km rail system is being built in various stages and will enable 60% of the city to be within 15 minutes walking distance of a station (Jain, 2008).

In Perth, Australia, a 172 km modern electric rail system has been built over the past 20 years, with stunning success in terms of patronage and the development of TODs; the newest section runs 80 km to the south and has attracted 50,000 passengers a day, whereas the bus system carried just 14,000 a day. The difference is that the train has a top speed of 130 km/h and averages 90 km/h, so the trip takes just 48 minutes instead of more than one hour by car. London, especially with its congestion tax which is recycled into the transit system, and Paris have both shown European leadership in managing the car (see Newman *et al.*, 2009).

While greening buildings, developing renewable fuel sources and creating more walkable communities are critical resilient infrastructure of the sustainable transport city, investing in viable, accessible transit systems is the most important component for them to become resilient to waning oil sources and to minimize the contribution of urban areas to climate change. Transit not just saves oil; it helps restructure a city so that it can begin the exponential reduction in oil and car use so necessary for a sustainable future and for human health (Matan *et al.*, 2012; Newman, 2006; Newman *et al.*, 2009; Newman and Matan, 2012).

2.2.7.3. *Street Planning and Mobility Management*

If cities build freeways, car dependence quickly follows. This is because the extra speed of freeways means that the city can quickly spread outwards into lower-density land uses as the freeway rapidly becomes the preferred option. If a city does not build freeways but prefers to emphasize transit, its streets can become an important part of the sustainable transport system. Streets can be designed to favor pedestrians and cyclists and wherever this is done, cities invariably become surprised at how much more attractive and business-friendly they become (Gehl and Gemzøe, 2000; Gehl *et al.*, 2006). The work of Danish Urban Designer Jan Gehl has shown how transport policies focused on creating a safe and vibrant pedestrian environment increase the economic, social and environmental benefits of a city center (Matan and Newman, 2012).

Sustainable mobility management is about 'streets, not roads'; the streets are used for a multiplicity of purposes, not just maximizing vehicle flow. The emphasis is on achieving efficiency by maximizing people movement, not car movement, and on achieving a high level of amenity and safety for all street users. This policy also picks up on the concept of integration of transport facilities as public space. One of the ways that United States and European cities are approaching this is through what are called 'complete streets' or, in the United Kingdom, 'naked streets'. This new movement aims to create streets where mobility is managed to favor public transport, walking and cycling, as well as lower speed traffic. The policy often includes removing all large signs for drivers which means they automatically slow down: in Kensington High Road in London the traffic accident rate has halved because of this.

Building freeways does not help create a resilient city. It will not help a city save fuel, as each lane rapidly fills, leading to similar levels of congestion that existed before the road was built (Nolan and Lem, 2001; Standing Advisory Committee on Trunk Road Assessment, 1994). Indeed, studies have shown that there is little benefit for cities in terms of congestion when they build freeways. There is no overall correlation between delay per driver and the number of lanes of major roads built per head of population for the 20 biggest cities in the United States (Urban Transportation Monitor, 1999).

Thus, for urban planners, the choices for a more sustainable city are quite stark, though politically they are much harder, as the allure of building more road capacity remains very high. Many cities that have confronted the provision of a freeway have been global leaders in the move towards more sustainable transportation. Copenhagen, Portland, Toronto, Vancouver and Zurich all had to face the cathartic experience of a controversial freeway. After a political confrontation, the freeway options were dropped. They decided instead to provide other greener options and build light rail lines. Cycleways, traffic calming and associated urban villages began to emerge. All these cities had citizen groups that pushed for a different, less car-oriented city and a political process was worked through to achieve their innovations. Similar movements are active in Australia (Newman and Kenworthy, 1999).

Freeways have blighted the centers of many cities and today there are cities that are trying to remove them. San Francisco removed the

Embarcadero Freeway from its waterfront district in the 1990s after the Loma Prieta earthquake. It took three ballots before consensus was reached, but the freeway has been rebuilt as a friendlier tree-lined boulevard involving pedestrian and cycle spaces. As in all cases where traffic capacity is reduced, the city has not found it difficult to ensure adequate transport, because most of the traffic just disappears. Regeneration of the land uses in the area has followed this change of transportation philosophy (Gordon, 2005).

What these projects have shown and encouraged is to "think of transportation as public space" (Burwell, 2005). With this changed approach to city planning, the small-scale systems of pedestrian movement and cycling become much more important. Pedestrian strategies enable each center in a city to give priority to the most fundamental of human interactions: the walking-based face-to-face contact that gives human life to a city and, in the process, reduces its ecological footprint.

Cycle-oriented strategies can be combined with the development of greenways that improve the green agenda and lower ecological footprint. Enough demonstrations now exist to show that pedestrian and bicycle strategies work dramatically to improve city economies and to integrate the green and brown agendas. Pedestrian and bicycle strategies in Copenhagen, most Australian cities, London, New York, San Francisco and Bogotá, as well as the dramatic changes in Paris with the Velib bicycle scheme and the growing awareness that it works in developing country cities as well, are all testament to this new approach to cities (Newman and Kenworthy, 2007).

2.2.8. *Urban Planning and Design for Green Urbanism Development*

The above seven resilient city types suggest that in order to create resilience in cities, there will need to be:

- Renewable energy strategies showing how to progressively tap local resources. Such strategies should involve recognition of renewable resources in and around a city as part of its capital base and establishing ordinances on buildings that facilitate the application of renewable energy.
- Carbon neutral strategies that can enforce energy efficiency, integrate with the renewable energy strategy and direct the carbon

offsets into the surrounding area. This can be enforced through planning schemes that mandate standards for significant reductions in carbon and water use in all developments that prevent the loss of arable and natural land, and direct planting to areas that are most in need of revegetation. Bioregional developments that can enable cities to partner with their surrounding rural areas in replanting and offsetting programmes will be needed.

- Distributed infrastructure strategies will be needed that enable small-scale energy, water and waste systems to flourish. This can be built into the requirements for urban development and can be encouraged by providing incentive packages with new buildings for such technologies as photovoltaic cells, grey water systems and water tanks, with local plans for the governance of community-based systems, as well as region-wide strategies for recycling sewage.
- Biophilic strategies that landscape buildings and facilitate water-sensitive urban design can enhance the thermal qualities of the city and help manage water and biodiversity better as well as provide recreation pursuits locally. This can be achieved through development controls that focus on how the rooftops (and walls) of buildings can be used for biophilic purposes, as well as zoning areas for urban biophilic activity, including growing biofuels, food and fibre, and biodiversity in and around the city.
- Eco-efficiency strategies that can 'close the loop' on a city's metabolism are needed. This will involve industrial ecology, linking industries to reduce their wastes, and assessing every step of urban development to achieve fundamental changes in the metabolism of cities. This can also be done by taking an audit of all the wastes of the city and seeing how they can be reused through stakeholder participation and government facilitation.
- Sense of place strategies are needed to ensure that the human dimension is driving all the other strategies. This can be assisted by local economic development strategies, by place-based engagement approaches to all planning and development processes and by the innovative use of 'sustainability credits', or complementary currencies, to implement local sustainability innovations as development bonuses. Iconic transformation of urban waterways is often a core activity in cities rediscovering their sense of place.
- Sustainable transport strategies incorporating: (i) quality transit down each main corridor that is faster than traffic; (ii) dense TODs

built around each station; (iii) pedestrian and bicycle strategies for each center and TOD, with cycle links across the city; (iv) plug-in infrastructure for electric vehicles as they emerge; (v) cycling and pedestrian infrastructure as part of all street planning; and (vi) a green wall growth boundary around the city preventing further urban encroachment.

2.3. Conclusion

Green urbanism for the city of the future is becoming an agenda that cannot be neglected as the global concerns accelerate over climate change, peak oil, water, waste, biodiversity and urban quality of life. Green urbanism offers ways of solving all these problems together. This is a challenge but it is also a great opportunity. There will need to be infrastructure to support the seven city types outlined if any city is to respond to these concerns. Examples have been provided of how each agenda is underway in some Western industrial cities; however, no city has begun to work equally on all seven areas. Eventually this will be required. Cities that are beginning to show leadership in the Asian region will thus be examined to see if there are enough encouraging signs for Asia to become the 21st century model for green urbanism.

CHAPTER THREE

The Renewable Energy City

3.1. Introduction

Renewable energy cities are those places where renewable energy is being taken seriously as the source of future urban energy. Cities can not only be consumers of fossil fuel energy but can also be a source of clean energy produced in the city and its region (Newman *et al.*, 2009). This chapter provides examples from cities and regions across Asia that are currently using renewable energy. The different projects show practices from the larger regional levels down to the small-scale building levels. A few of the examples are still proposals and are still being implemented.

The developed world has been the pioneer in the technology of harnessing the sun's radiation and resulting energy in photovoltaic (PV) and wind power. But Asian countries have now taken over as the source of renewable energy. Initially, solar panels were installed in rural areas in many developing countries across Asia where electricity is scarce. However, now, PV cells are being used to power many households, towns and cities within the region in an effort to decrease fossil fuel consumption.There are a number of proposed cities to be developed in Asia that incorporate the use of solar power as their main source of electricity, referred to as 'Solar Cities'.

In 2004, investment in renewable energy was mainly in the developed world with Bloomberg Energy showing that renewable investment was $15 billion in the developed world to $4 billion in the developing world (Figure 3.1). However, by 2011 this had crossed over with $70 billion being invested in renewables in the developed world with $72 billion in renewables in the developing world (United Nations

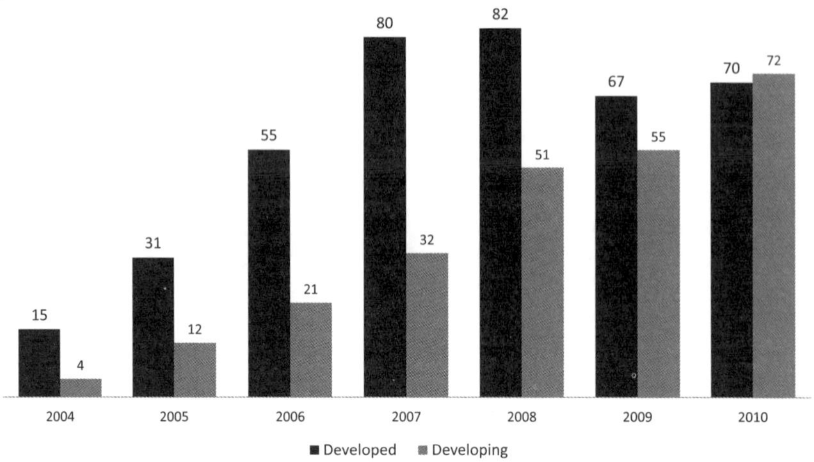

Figure 3.1 New Financial Investments in Renewable Energy for the Developed and Developing Countries, 2004–2010.

Note: The new investment volume adjusts for re-invested equity and total values include estimates for undisclosed deals.

Source: Redrawn from United Nations Environment Programme and Bloomberg New Energy Finance, Figure 5, 2011.

Environment Programme and Bloomberg New Energy Finance, 2011). This is mostly in India and China though as shown below there are many others beginning this transition.

The renewable energy journey in China is an amazing story. In 2005, China enacted the landmark Renewable Energy Law promoting the use of wind, solar, hydro, biomass, geothermal, tidal and other non-fossil energies, in line with the country's advocacy of green technology. Implementation of this policy started in 2006, tapping into the renewable energy market in China, setting targets to reduce the country's carbon emissions as well as increase the share of renewable energy.

In 2006, China initially set a projected target of an 8% share of primary energy to come from renewable sources by the end of 2020. However, with the positive progress of this clean energy industry, they have now increased their target to 15% by 2020, amounting to 500 GW of energy coming from renewable sources (United Nations and Asian Development Bank, 2012, p. 70, Box 3.13). Furthermore, the Renewable Energy Law outlines a commitment to invest $180 billion to accomplish these energy goals. In 2008, China was the "largest producer of clean-tech in monetary terms, earning more than 44 billion euros, or 1.4% of its GDP, leading the world in many clean-tech sectors, including wind turbines, solar photovoltaic hardware, and high-speed rail technologies" (United Nations and Asian Development Bank, 2012, p. 53). Additionally, they have set targets for reducing carbon intensity by 40–45% by 2020 across five major cities (Beijing, Tianjin, Shanghai, Chongqing, and Shenzhen) and two provinces (Hubei and Guangdong) (Webster, 2012).

In 2009, the total renewable energy capacity in China had already reached 226 GW. Through this policy, China's renewable energy production has seen a massive boom over recent years. The country is now the largest wind turbine manufacturer and the largest manufacturer of solar PV, with its solar PV production accounting for 40% worldwide. The big reductions in the cost of PV worldwide are mostly due to the efficiencies in production made in China.

To give a glimpse of what China is planning in the near future, the government's current draft plan targets 300 GW of hydropower, 150 GW of wind power, 30 GW of biomass power, and 20 GW of solar power by 2020. The total of 500 GW of renewable energy capacity is already one third of the total 1600 GW of projected power capacity of China by 2020 (Martinot and Li, 2010). In order to achieve this, most new investment is in renewable energy as is happening across the world (see Figure 3.2); in 2011, 86% of the world's new investment in power was in renewable energy and most of this is in PVs due to the dramatic reduction in the cost of production in China (Newman, 2012).

The National Development and Reform Commission (NDRC) in China announced a program for five low-carbon pilot provinces and eight low-carbon pilot cities in 2010. The provinces are Guangdong, Liaoning, Hubei, Shaanxi and Yunnan; the cities are Tianjin, Chongqing, Shenzhen, Xiamen, Nanchang, Guiyang, Baoding and Hangzhou. This program covers greenhouse gas accounting, low-carbon development planning, industrial and economic policies, government official training, communications and international cooperation. Along with low-carbon industry technology, this plan also encourages low-carbon lifestyles and a strong focus on renewable energy (Karlenzig and Zhu, 2010; United Nations, Bureau International des Exhibitions and Municipal Government of Shanghai, 2011). An example of this is the city of Baoding. The city has had a strong economic growth rate, which is attributed to a 40% growth in companies that produce and deploy

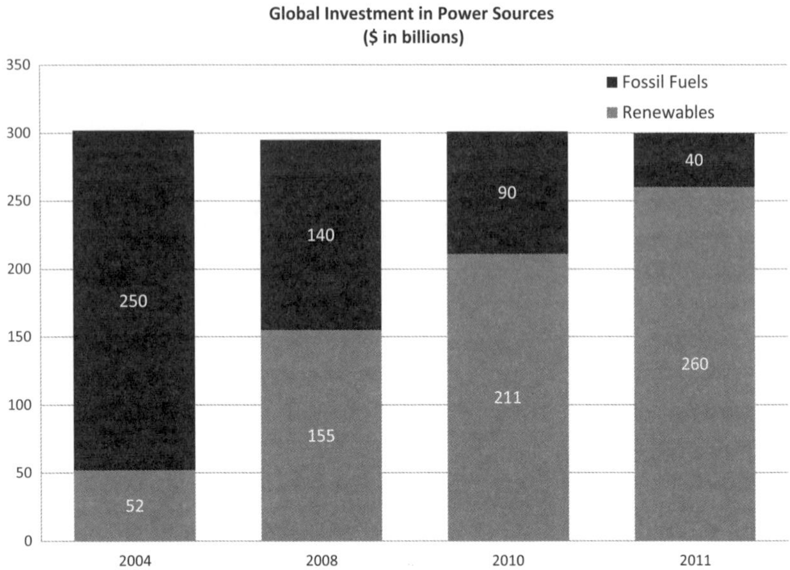

Figure 3.2 Investment in Renewable Energy.

Source: Newman, 2012.

low-carbon technologies (particularly wind, photovoltaic, thermal, solar, biomass and energy-efficient technologies). This focus started in 1992 with the establishment of the Baoding Industry High-tech Development Zone, when the city made a conscious effort (led by the mayor) not to pursue traditional industries, rather to pursue industries that produce low-carbon technology. This has resulted in 20,000 jobs in clean energy technologies (in 2010).

The illustrations of renewable energy cities will be divided into those with an emphasis on PVs, followed by those emphasizing wind energy and finally those emphasizing geothermal energy.

3.2. Case Studies of Renewable Energy Cities in Asia: Photovoltaics

3.2.1. *Kaohsiung, Taiwan*

The City of Kaohsiung, located in southwestern Taiwan, has been doing much to increase its sustainability, particularly in the way of increasing public transport, modernization of the sewage system, improving the quality of potable water, diversifying their sources of water as well as promoting the reduction of CO_2 emissions and the production of solar energy (Her, 2008; Republic of China (Taiwan), 2010; Taiwan Sustainable Cities, 2011).

Kaohsiung, with a population of 2.9 million people, is the second largest city in Taiwan. During the 20th century, Kaohsiung experienced rapid industrialization and urbanization, becoming a major location for industries. By the 1990s, Kaohsiung had over 6,000 factories. This rapid industrialization, however, resulted in much pollution. The city has been trying to reduce this through the use of clean energy sources.

In June 2009, the Taiwanese government passed the Renewable Energy Development Act with the purpose of adding 6,500 to 10,000 MW of energy from renewable sources within the next 20 years and to also make industries reduce their CO_2 emissions through offsets or by adopting cleaner energies. The Act aims to return the county's greenhouse gas emissions to pre-2000 levels. In 2011, Taiwan generated approximately 2,278 MW from renewable sources, accounting for 5.8% of its electricity consumption. Kaohsiung's government agencies have invested over NT $37 billion in a 'Green Energy Industry Plan', aimed at increasing the city's green energy industrial value to NT $1.158 trillion by 2015. This plan identifies the Kaohsiung Science Park as a 'green energy cluster' designated for promoting LED, biotech, telecommunications and solar businesses.

The Kaohsiung municipality has a high rate of sunshine, making it a prime location for the production of solar energy. In February 2010, the largest high-concentration photovoltaic (HCPV) solar power plant in Asia began operation in Lujhu, located in the Greater Kaohsiung area. The plant was funded by the Institute of Nuclear Energy Research and can produce 1 MW of energy, with potential to be upgraded to 10 MW. The majority of the solar cells and components, including the solar concentrator modules, the solar trackers, the solar-electric system and the central monitoring system were developed by domestic companies. The Chinese/Australian company Suntech Power Holdings Co. Ltd. announced a 4.7 MW solar plant to be located in the greater Kaohsiung area in 2010. This would be Taiwan's largest solar power plant and would double Taiwan's solar capacity once completed. Furthermore, the Kaohsiung government has been providing subsidies for solar hot water systems. There are also plans to build a solar energy industry zone along the harbor as part of a land reclamation project.

A methane gas plant in Kaohsiung, the Hsichingpu methane plant, harnesses methane gas from a landfill underneath the Metropolitan Park. This plant supplies electricity to 4,600 households. It began operation in 2000, and is estimated to reduce 5,000 tons of greenhouse gas emissions per year over its 20 years life span (Taiwan Sustainable Cities, 2011).

Kaohsiung has made other remarkable strides at becoming a healthier and more sustainable city which are summarized in Box 3.1. Part of this transformation has been as an economic vehicle to attract investment, illustrating that the future of urban economies will be based on the technologies of green urbanism. All cities will compete for future economic advantages with green urbanism and so need to use each opportunity for development to be a trial of these green urbanism technologies. Box 3.2 shows how Kaosiung has done this with its iconic new stadium and its solar canopy.

Box 3.1 A Healthier and More Sustainable Kaohsiung: Other Actions

Transport

- To help to increase the mode share of public transport (which in 2006 was 4.3%), the city has been focusing on providing new options and on upgrading the efficiency of the existing services. In 2008, the city has developed a Mass Rapid Transit (MRT) system: the Kaohsiung MRT (KMRT). The system has 2 lines, 36 stations and has 42.7 km of track. Furthermore, the plan is to expand the existing transport system's connections by adding hundreds of buses to the road, including 100 low-floor buses for wheelchair passengers, and doubling the shuttle bus connections to the KMRT line. The public bus fleet is fueled by a biofuel and petrol blend. Six hydrogen fuel buses are being tested.
- The city has been proactive at encouraging people to travel by bicycle rather than by motorcycle, introducing 150 km of cycle paths and a bike rental scheme called C-Bike, which includes 4,500 bikes at 50 rental locations.

(*Continued*)

Box 3.1 (*Continued*)

- The city of Kaohsiung is also focusing on improving fuel efficiency both in the city's municipal vehicle fleet and in the private vehicle and motorcycle fleet, in part focusing on alternative fuels. The city has converted all of its garbage trucks to biodiesel fuel in 2006, and is offering subsidies towards the purchase of cleaner vehicles and motorcycles.

Sewage

- The city has been very proactive at upgrading their sewage treatment system and connectivity. From 1979 to 2001, the city upgraded the sewage treatment facility. In 1995, only 1.1% of households had a municipal sewage system; in 2010, 61% were connected to the system.

Water

The city has been actively trying to improve the quality of its tap water and has been very proactive at exploring water recycling:

- The city's first industrial wastewater reclamation plant opened in 2011. It serves the Nanzih Export Processing Zone. Initial estimates of the plant maintain that it will produce 1,800 tonnes of H_2O/day, with the plan of upgrading to enable a capacity of 100,000 tonnes of H_2O/day.
- The city also encourages water recycling at a business scale. China Steel's Kaohsiung plant requires 165,000 tonnes of H_2O/day (55% of the entire city's industrial water needs). They have installed an industrial wastewater recycling plant that will reuse up to 13,500 additional tonnes of H_2O/day.

Energy from the Ocean

- In 2008, the Bureau of Energy (BOE) established a three-year program researching the possibility of producing energy using ocean thermal energy conversion that exploits the temperature differences between the shallow and deep waters and uses this to generate power. Researchers have determined that the thermal energy in the waters off Taiwan's east coast could produce 2% of the country's power requirements (Industrial Technology Research Institute, 2011).

(*Continued*)

Box 3.1 (*Continued*)

- The city is also looking into energy produced by wave power.

Love River

- Love River, which runs through the heart of the city, was once the major dumping ground for the city's runoff pollution. Since 1979, the city has been improving its sewage treatment, re-vegetating the area, and adding public areas, bike paths and public art spaces along the river bank. The river now is home to over 50 aquatic species. The city runs five solar-powered tour boats along the river.

Green Spaces

- The city of Kaohsiung is also focusing on increasing the provision of green spaces and has provided over 750 hectares of parks, fields and a wetland ecological corridor.
- Kaohsiung has eight wetlands that form a pathway running from the north to the south of the city comprising over 50 million m^2.

Reducing Meat Consumption

- In an effort to reduce demand-side CO_2 emissions and to teach environmental awareness, the Kaohsiung government encourages citizens to adopt a vegetarian diet for a day per week. Many schools and universities now have a vegetarian day, with 33 schools having two vegetarian days per week.

Sources: Bureau of Energy, n.d.; Her, 2008; Industrial Technology Research Institute, 2011; Kaohsiung City Government, 2011; Republic of China (Taiwan), 2010; Taiwan Sustainable Cities, 2011.

Box 3.2 Kaohsiung Solar Stadium, Taiwan

The Kaohsiung solar stadium, nicknamed the 'flying dragon', was designed by Toyo Ito and is shaped liked a dragon or a 'C' and has a solar canopy that provides weather protection. It seats 50,000 people and occupies 14,155 m^2. The stadium was finished in 2009. The stadium cost about $150 million to construct and used recyclable

(*Continued*)

Box 3.2 (*Continued*)

materials that were all produced in Taiwan. The stadium is clad with 8,844 solar panels and produces a staggering 1.14 million kW of electricity a year. This produces all the electricity for the stadium's 3,300 lights and two screens. The solar panels were all produced by Lucky Power Technology Company. During the periods when the stadium is not in full use, the excess power produced is sold to the main electricity grid. This excess power provides electricity to approximately 80% of the neighboring area's energy needs. The stadium is predicted to produce 660 less tons of CO_2 per year compared to a comparable stadium that is traditionally powered.

Source: Gizmotrix, 2009.

Sources: Industrial Technology Research Institute, 2011; Renewable Power News, 2009; Taiwan Sustainable Cities, 2011.

3.2.2. *Dezhou "Solar Valley", China*

In line with China's renewable energy policy, Himin Solar Energy Group has invested in the China Solar Valley in Dezhou which is envisioned to be "the biggest solar production base in the world". In partnership with the Dezhou local government, the China Solar Valley project aims to promote China's solar industry. Situated in over 330 hectares of land area, the China Solar Valley is dedicated to the PV industry and is a center for manufacturing, research, tourism, education, and promotion of solar energy technologies. The Solar Valley is home to various facilities, including:

- The Solar Research Institute Center — a joint research scientific institution between the domestic China Academy of Science and various universities.

- A Solar Testing Center — a center aimed at being one of the most advanced and comprehensive solar testing laboratories in the world.
- A Science and Technology Museum — the first Solar Arts Center in the world.
- An International Conference Center — this center hosted the Fourth World Solar City in 2010.
- A Micro-emission Hotel.
- A Renewable Energy College and University.
- A Tourist Center and Solar Energy Amusement Park.
- Solar Homes and Pavilions.
- Solar Condominium Residences.
- A Solar Stadium.
- PV Production Facilities.

The local government's construction authority stipulates that all new buildings are now required to be equipped with solar water heaters and more than 80% currently have them installed. The Sun-Moon Mansion, shown in Figure 3.3, is a main architectural feature of the Solar Valley and is home to Himin Solar Energy Group's headquarters. This

Figure 3.3 The Sun-Moon Mansion — A Good Example of a Building Mainly Powered by Solar Energy.

Source: Yoneda, 2011.

is considered to be the world's largest solar structure. With a total building area of 75,000 m^2, it showcases green technology at its finest. Integrated with over 2,000 sun collectors, solar water heaters, cooling equipment and ceiling radiators, the building accumulates as much as 88% of energy savings amounting to annual savings of 6.6 kWh of electricity and annual pollution reduction of 8,672 tons of CO_2. The city's main boulevards are now also lined with solar-powered street lamps.

The Solar City project has helped the economy of Dezhou through investments in the solar industry. The site is now home to 100 solar enterprises with 586 solar product patents, 5 solar scientific research programmes that have also provided 800,000 jobs for the local people in Dezhou. The Solar Valley became the host of the Fourth International Solar Cities World Congress in 2010 following the steps of the previous host cities of Taegu in South Korea, Oxford in the United Kingdom and Adelaide in Australia. The builders of this community are hoping that more cities in the future will follow in their steps as they have proved that a development centered on the use of renewable energy can be successful. Other Asian countries like Japan, South Korea, and Taiwan are now cooperating with Himin Solar Energy Group in furthering the solar industry in this city. This ambitious project aims to demonstrate that China is leading the global clean energy industry (Higgins, 2010; Himin Solar, 2012; Huang, 2009; International Solar Cities Initiative, 2010; Xu, 2010; Yoneda, 2011).

3.2.3. *Rizhao City, China*

Rizhao City, home to 3 million residents, is known as a sunshine city. It is located in the southeastern Shandong province along the coast and has an abundance of natural resources. Rizhao City has a unique planning emphasis focused on innovation, public education, science and technology. As part of this, the City has had a focus on large-scale and small-scale renewable energy (C40 Cities, 2011; Clean Energy Awards, 2007; Future Policy, 2007; Levesque, 2007; Sustainable Cities, n.d.; World Future Council, n.d.).

The City has many policies aimed at encouraging the use of clean energy, limiting wasted energy, improving energy efficiency and reducing greenhouse gas emissions. The government provides residential subsidies, consumer environmental awareness programs and exhibitions of solar energy products. Numerous solar products are produced in the City, including solar cookers, solar heating, and solar cell technology. Furthermore, the City has developed a new technology that is able to produce electricity from marsh gas (waste-derived methane) using wastewater. The Rizhao Municipal Government provides capital support from an environment-protection capital fund to companies that use wastewater to produce marsh gas for electricity. The companies treat the wastewater to discharging standards and are then excused from paying fees related to discharging, water sourcing and treatment. This policy reduces the pollution discharged through wastewater and enables more electricity to be produced by marsh gas. In rural communities, farmers receive subsidies for utilizing marsh gas energy, with 15,000 rural households utilizing this form of energy. Marsh gas production has an annual generating capacity of 90 million kWh in this City.

Approximately 1.5 million people directly utilize or benefit from the solar energy, the marsh gas energy and various energy efficiency measures. In 2007, 400,000 households used solar energy and 15,000 households used marsh gas within Rizhao City (Figure 3.4). The City saves an estimated 3.8 billion kWh of electricity per year through the utilization of solar energy in agriculture, construction, lighting and heating. Rizhao City has experienced significant improvements in air quality as a result of this policy, particularly through the reduction of SO_2 and CO_2 emissions. The City estimates that through the use of solar energy and marsh gas, CO_2 emissions could be reduced by 3,250,000 tons, SO_2 by 21,300 tons, and ash dust by 20,000 tons.

The changes in Rizhao City have been a combination of three factors: government policy focused on clean energy production, business interests and strong leadership within the City Government. The Municipal Government provides public activities aimed at promoting energy efficiency. These include lectures, publications and on-site guidance. These programs increase the awareness of the importance of energy efficiency and the need to use clean energy. The City of Rizhao has had strong leadership facilitating the adoption of new energy supply systems and providing support and demonstration projects, including installing solar on government buildings and in employees' homes.

Figure 3.4 Rizhao City Solar Installations.
Source: Clean Energy Awards, 2007.

3.2.4. *Sakai City, Japan*

In 2007, the City of Sakai announced the 'Sakai City Waterfront Mega Solar Power Generation Plan' for a 20 hectare industrial landfill site to be turned into Japan's first mega solar power plant and would be a collaboration between the City of Sakai, Sharp Japan and Kansai Electric Power. The plan was for two large solar power generation plants that would have a combined output of 38,000 kW, with 10,000 kW from one of the plants and 28,000 kW from the other with the 10,000 kW plant starting commercial operations in 2011. The plants utilize 70,000 silicon thin film panels produced by Sharp, illustrated in Figure 3.5. These plants would reduce CO_2 emissions by 10,000 tons annually.

The construction phase of the solar power plant adopted efficient construction methods. The installation of the solar panels requires 150,000 pieces of concrete foundation. This was constructed at a dedicated centralized unit to facilitate the construction process. Furthermore, efficiencies in construction were enabled through a roller rail transporting belt set along the solar panel line. The construction period was further shortened by having the solar panels pre-adjusted to the 15-degree angle ready for mounting. Typically the solar panel would be mounted at a 30-degree angle, which is ideal to catch the sunlight;

Figure 3.5 Sakai Solar Power Plant Uses 70,000 Silicon Thin Film Solar Panels Made by Sharp.

Source: Seagate, 2008.

however, for this installation, it is mounted at a 15-degree angle, which, based on recent findings by a research agency in Gobo, Wakayama Prefecture, can efficiently catch the sunlight. The 15-degree angle setting eliminates the requirement of a standard angle frame, allowing the solar panel to be directly mounted on the concrete foundation, resulting in time and cost savings.

3.2.5. *Pal Town, Ota City, Japan*

Ota City is 50 miles northwest of Tokyo in Gunma Prefecture and is aiming to be the world's first fully solar-powered city (although it is not the only city aiming to be this). Starting in 2002, the government funded initiatives aimed to test whether the use of solar power could minimize blackouts. The study cost 9.7 billion *yen*, with the government providing free solar panels to the majority of households within the neighborhood of Pal Town. Pal Town was chosen as the test site for the solar panel experiment because of its sunny location and availability of space.

This test has been very popular with many people moving to the 'Solar City' in the past few years. Three-quarters of Pal Town's homes have been covered by solar panels and many small solar power generators are connected to the power grid. One of the major attractions for residents has been the ability to minimize their power bills (Sharp, 2008; Solar Power — PV Panels, 2012).

3.2.6. *Masdar City, Abu Dhabi, United Arab Emirates*

Masdar City is a planned car-free, sustainable, zero-carbon and zero-waste city located 17 km southeast of Abu Dhabi in the United Arab Emirates (UAE). Once completed it will be one of the largest and most advanced sustainable cities in the world. Masdar was designed by the British architectural firm Foster and Partners and its construction was initiated in 2006. Masdar City occupies 6 km^2 and uses designs based on traditional Arabic forms of architecture. The city will cost an estimated US$22 billion and originally was planned to be built in approximately eight years, with construction planned in seven phases, the first of which was a large solar plant that provides power for the construction of the city (Figure 3.6). The city was due to be completed by 2016 though this is now delayed due to the global financial crisis (GFC); it will be home to 100,000 people and 1,500 jobs in green urban technologies. The city is part of an ambitious plan by the government of Abu Dhabi and

Figure 3.6 Photovoltaic Arrays in Masdar City.

Source: Design Scene, 2009.

photovoltaic company Masdar PV to become one of the largest solar technology companies in the world (Design Scene, 2009; Oppenheimer, 2010; Pohl, 2009; Singh, 2011; Stanton, 2011).

The design of Masdar City combines traditional Arabic planning principles and innovative energy, water and waste technologies as well as sustainable transport. The whole city is a compact, walled city centered on shaded walkways and narrow streets, many of which will be only 10 feet wide, designed to create an environment suitable for the extreme desert climate following traditional Arab design principles. Some streets will also feature canals to further cool the climate. The city is car-free, with the narrow shaded streets designed to encourage walking. Public transport, through the use of a fully automated electric personal rapid transit system, is provided for further distances and they will have public transport stops every 200 m. A light rail transport system will eventually link Masdar to the airport and to Abu Dhabi. The fossil fuel consumption, water use and waste production are all predicted to be much less than a conventional city, with fossil fuel use predicted to be 75% less, water demand 300% less and waste production 400% less.

The main source of renewable energy will be solar; however this will be complemented by wind, biomass and geothermal energy. There are two large solar plants (the first is built) along with solar roofs which provide both shade and energy. Masdar PV's projects include a large polysilicon factory in Abu Dhabi, which once finished will provide feedstock for additional manufacturing activities like that of manufacturing PV cells and modules. The area will also produce biofuel from palm and mangrove plantations that will surround the city, providing a green belt and also the raw materials for the industry. Solar thermal energy will be used to heat water and to cool buildings. Biomass energy will be produced from the waste.

Water will be provided through desalination, rainwater and dew collection. The desalination plant will be solar-powered. Water will also be recycled to help offset some of the water demand. The water demand of Masdar is predicted to be about 60% lower than that of a comparably sized city.

The built form of Masdar is primarily low-rise (5 to 8 stories) with narrow streets and the buildings are designed to have low energy requirements and to be able to run entirely on renewable energy. The buildings feature highly energy-efficient passive solar design and allow for natural ventilation. There are wind towers that utilize natural convection to provide ventilation. The narrow streets provide natural cooling through trees and overhead PV arrays that act as shading devices. From these features, it is predicted that the power needs will only be about a quarter of that of a similarly sized city.

Some of the other sustainable features of Masdar include:

- Water features where the water is stored under the ground and can be triggered by the movement of pedestrians.
- An interactive lamp, inspired by the Oasis fire. This is also the forum of a three-dimensional interactive media installation.
- Interactive heat-sensitive lighting activated in response to pedestrian traffic and the use of mobile phones.
- Roof gardens that allow space for food production, power generation, water use efficiency and the reuse of organic food waste.

There is, however, some controversial elements to the design. In particular, the design of the public transport system has been controversial. The plan is for a system of magnetic solar-powered individual vehicles, referred to as pod cars or personal rapid transport which have caused many teething problems and have questionable efficiency in moving large numbers of people.

3.2.7. *Donggwang, Jeju Island, Korea*

Donggwang is a small village located on the western half of Jeju-do, South Korea. The village receives all of its power from solar energy. The installation of the solar system was subsidized by the government in 2004 with 70% of the installation fees for the 40 houses and one school in the village paid for. Most roofs are installed with a 2 kW solar system. The village motto is 'clean city — clean island' and through the use of solar energy the village has achieved energy independence (Hudson, 2008b; Thomas, 2008; World Wind Energy Association, 2006).

3.2.8. *Sanyo Solar Ark, Japan*

The Sanyo Solar Ark is a photovoltaic power generation facility located in the Gifu prefecture, central Japan (Figure 3.7). The facility has over 5,000 solar panels and generates 530,000 kWh per annum. The facility was designed to represent an ark moving towards the 21st century (like Noah's ark), and is 315 m wide (1033 feet), 37 m tall (121 feet) and is

Figure 3.7 Sanyo's Solar Ark.

Source: Inhabitat, 2008.

constructed out of steel. The Ark is centered on four pillars and uses single-crystal silicon solar cell modules. The Solar Ark has a strong educational component, featuring a museum of solar energy, the Solar Lab. The Ark is surrounded by water fountains and ponds and has a pair of solar wings providing shade over the entrance. The Ark is completely solar-powered and sells electricity which provides revenue for Sanyo's environmental fund. This fund supports environmental conservation organizations (Kriscenski, 2008; Sanyo, 2012).

3.2.9. *Thyagaraj Stadium, New Delhi, India*

The 16,000 m^2 Thyagaraj Stadium in Delhi is India's first eco-friendly sports complex (Banerjee, 2009; Banga, 2010). It is located in the Indian National Army colony in South-Central Delhi. The rectangular stadium was designed by Australian firm Peddle Thorp and features extensive solar infrastructure (Figure 3.8). The solar capacity of the stadium generates around 1.4 million units of electricity per year, providing for the electricity needs of the stadium and providing 11 kW excess power to the electricity grid. The solar electricity production of the stadium is estimated to reduce emissions equivalent to 1,200 tonnes of CO_2 per year.

The stadium utilized sustainable construction methods, as well as sustainable building features and facilities to reduce water consumption. The stadium was constructed using fly ash bricks with the walls containing a cavity that provides an acoustic and thermal barrier; it features specialized double-glazed glass that allows for a high amount

Figure 3.8 Solar Panels on the Rooftop of Thyagraj Stadium, New Delhi.

Source: Suniva Inc., 2010.

of light but only a low level of heat to enter the building. Water conservation measures include the use of recycled water and rainwater harvesting systems. The facility also produces electricity through a wind turbine, which generates 3.5 MW of power.

3.2.10. *Solar Steam Cooking at Tirumala, Andhra Pradesh, India*

Tirumala Tirupathi Devasthanam (TTD) in Andhra Pradesh is a very popular pilgrimage destination where people come to receive the blessings of Lord Tirupati Balaji. The temple authorities provided food to devotees every day at the temple, whilst facing fuel and electricity shortages. In order to overcome this problem, in 2002 a large solar steam cooking system able to cook food for 15,000 people per day was installed (Figure 3.9). The system cost Rs. 110 million and was installed by M/s Gadhia Solar Energy Systems, Valsad, under a demonstration scheme of the Ministry of Non-conventional Energy Sources (MNES) with 50% financial support. The rest of the cost was borne by the TTD trust. The solar steam cooker uses an automatic tracking solar dish concentrator that convert water into

Figure 3.9 World's Largest Solar Steam Cooking in Tirumala Tirupathi Devasthanam.

Source: Fractal Enlightenment, 2008.

high-pressure steam. This steam is then able to be utilized for cooking. The system can produce over 4000 kg of 180°C steam per day. 10 kg/cm^2 is adequate to cook two meals for approximately 15,000 persons. The system saves approximately 118,000 litres of diesel per year, equated to Rs. 2.3 million (Government of India, Ministry of New and Renewable Energy, 2002–2003; Muthu and Raman, 2006; Solar Energy Demystified, 2010).

This solar steam cooker has been so successful that six other similar cooking systems have been established in India. These types of cookers would be very useful in places where rice is the staple food and cooking needs to be done on a very large scale and is a good example of green urbanism being applied in a place-sensitive way.

3.2.11. *Zero Energy Building, Singapore*

The use of photovoltaics is rapidly growing in Singapore (from 50 kW in 2007 to 2000 kW in 2009) with government demonstration programs worth S$68 million being applied to commercial projects and residential projects. There are a number of exemplary projects including the Building and Construction Authority's (BCA) Zero Energy Building (ZEB) and 30 solar precincts planned by 2015.

ZEB is a retrofitted three-story office building along Braddell Road (BCA Green Mark, n.d.; Wang, 2009). The S$11 million [US$7.9 million] retrofit was designed to enable the building to produce energy equal to what it consumes and included a redesign of the façade and roof, and the implementation of a monitoring and evaluation (M&E) system. ZEB features offices, a library and classrooms, all of which are used to experiment with innovative building designs and energy-efficient building solutions. The building utilizes solar panels installed on the roof and in other appropriate areas to generate 207,000 kWh of electricity per year. The solar panels cover 1,300 m^2. The building only needs three hours of sunlight to supply all of the energy requirements needed for a day. ZEB also features a green wall, which helps to reduce temperatures (external wall temperatures reduced by up to 12°C), and utilizes a solar chimney which removes warm air, providing natural cooling. The building utilizes natural light and natural shading, including through light pipes (Figure 3.10). ZEB is also fitted with a single coil twin fan ventilation system providing fresh air. This system is able to control

Figure 3.10 PV Arrays and Light Pipes that Channel Natural Light into the Upper Floors of BCA's Zero Energy Building.

Notes: The dome protects mirrors that redirect light to a distance of 5 m into the building. In the background are solar chimneys used to cool a sports hall within the building.

Source: The Star, 2009.

the flow of fresh and recycled air separately throughout the building, enabling fresh air to be channeled into the rooms that are occupied and that need air. The energy use of ZEB is 40% to 50% more efficient than a comparable office building.

The government of Singapore is actively trying to encourage more sustainable retrofits of the existing building stock, providing monetary incentives under the Green Mark certification scheme. The aim is to get 80% of all building stock to be Green Mark-certified by 2030.

3.3. Case Studies of Renewable Energy Cities in Asia: Wind Power

Wind power has been harnessed for hundreds of years but has been rebuilt with new technology to provide electricity at small-scale and large-scale facilities. A number of Asian countries are now operating wind farms after recognizing the great potential of wind energy and its numerous advantages, such as facilitating the supply of electricity in remote areas and helping in the reduction of greenhouse gases. Wind power is also rapidly approaching grid parity in its pricing thus it is becoming mainstreamed (Bloomberg New Energy Finance, 2012).

3.3.1. *Jeju Island, Korea*

Jeju Island, Korea, has a large wind farm and abundant wind resources for power generation. The plant is planned to have a power generation capacity of 500 MW by 2020. This would cover 20% of the island's energy consumption (Hudson, 2008a).

3.3.2. *Tamil Nadu, India*

India's current power needs come mostly from coal. The country has a dysfunctional power production and distribution system which experiences many blackouts due in part to over-consumption of power. In order to sustain growth and be able to provide electricity to the populace, alternative sources of electricity are needed. Wind energy is increasingly being sourced to provide renewable energy (Purohit and Michaelowa, 2007). India ranked fifth in the world for wind power generation in 2011 with a capacity of 14989.40 MW accounting for 8% of India's power capacity (Indian Power Sector, 2011).

The area of Tamil Nadu produces 5900 MW of wind energy (2011). This was enabled by the technical innovations in wind turbine systems made primarily by companies such as Suzlon and Vestas. The Muppandal wind farm, in Aralvaimozhi, is the largest in the subcontinent, providing a showcase for India's $2 billion clean energy program (Lee, 2010a). This program provides tax incentives to foreign companies to develop wind farms. Furthermore, the area offers competitive tariff rates. The Tamil Nadu Electricity Regulatory Commission (TNERC) regulates that the state's power distribution companies have to buy a specified minimum percentage of their power from renewable energy sources (Chadha, 2010). It has unveiled many financial incentives (such as Special Capital Subsidy) for the promotion of solar energy and other renewable energy sources. For instance, under the Micro, Small and Medium Enterprises Policy 2008, companies can avail a subsidy of up to Rs. 3 million for the procurement of solar energy equipment. Tamil Nadu Electricity Board is required to buy at least 14% of the power generated from renewable energy.

3.3.3. *China's Wind Power*

China has abundant wind power potential. The China Meteorology Research Institute maintains that the exploitable wind energy nationwide is 1,000 GW, with around 250 GW available on land and 750 GW in offshore wind energy. China began investing in commercial-scale wind power in the 1990s. The single-unit capacity has increased from 100 kW, 200 kW, and 300 kW to 600 kW, 750 kW, and 1,300 kW over the intervening years (China Wind Power Center, 2012).

The Chinese government established a goal of having 30 GW of wind power by 2020. In order to further this, the government issued a circular on 'Questions Concerning Advancement of Renewable Energy Resources' in 1999. This circular contains the preferential policy for renewable energy. The policy provides direction on the financing of wind power projects, setting a fiscal subsidiary of 2% on projects requiring a bank loan and for preferential treatment of 5% on return rate of investment for wind power projects using Chinese-made equipment by the top three Chinese wind-turbine manufacturers: Xinjiang Goldwind, Sinovel and Dongfang. In addition, China's National Development and Reform Commission (NDRC) require that large power-generating

Figure 3.11 Inner Mongolia Wind Farm, the Biggest Wind Power Generating Base in China.

Source: Reuters and Gray, 2010.

companies incorporate at least 3% of wind energy by 2010 and 8% by 2020.

The majority of China's wind farms are located in the north, accounting for 70% of the installed wind power, and along the coastal areas of Shandong, Fujian and Guangdong. Much of the installed capacity, one quarter of China's wind power production, is located in inner Mongolia (Figure 3.11). The inner Mongolian wind farm has a generating capacity of 100 MW, making it the biggest wind power generating base in China.

3.3.4. *Bangui Windmills, Ilocos Norte, Philippines*

The Northwind Bangui Bay project is located in Bangui, Ilocos Norte, Philippines. Built by Northwind Development Corporation, the project aims to help reduce greenhouse gas emissions and to accelerate rural electrification in the region. The project is considered to be the first wind farm in the Philippines and to be the biggest in Southeast Asia. Started in June 2005, the farm has 15 wind turbines located along the shore of the South China Sea placed 326 m apart (Figure 3.12). Each turbine produces up to 1.65 MW for a total of 24.75 MW for the plant. Originally the Northwind Bangui Bay Wind Farm provided 40% of

Figure 3.12 Wind Turbines along the Shores of Bangui, Ilocos Norte.

Source: Waypoints, 2007.

the power requirements in the Ilocos Region. Phase 2 of the farm was completed in 2008. This upgrade was targeted to increase the energy production of the wind farm to enable it to supply the region with as much as 70% of its power requirements.

The Northwind Bangui Bay project superseded another electricity proposal for the area — a coal plant. This previous plan seemed to be the more practical and cheaper option in terms of project cost. However, surprisingly, the majority of the local residents did not approve the coal-powered electricity plant proposal as it did not guarantee stable pricing and because it would certainly have negative impacts to the environment. With the people's awareness and concern for their environment, the coal plant did not materialize and instead, the wind farm was approved.

It is estimated that the Northwind Bangui Bay project saves 56,788 tons of CO_2 equivalents of greenhouse gases per year. The project also resulted in a 5% cost savings for electricity consumers. The wind farm became the first Philippine recipient of the Carbon Emission Reduction Certificate (CER) from the United Nations Framework Convention on Climate Change (League of Provinces of the Philippines, n.d.). With the success of this project, there are now plans of establishing more wind farms throughout the country.

The Renewable Energy Act was established in the Philipines in 2008, and is considered to be the most comprehensive renewable energy law in Southeast Asia. The act includes a net metering scheme where consumers can sell power to the grid at an approved feed-in tariff and buy power as necessary at the normal retail tariff. The feed-in tariff provides a guaranteed fixed price for at least 12 years for electricity produced from emerging renewable resources such as wind, solar power, ocean, run-of-river hydropower and biomass (Congress of the Philippines, 2008, as cited in United Nations and Asian Development Bank, 2012, p. 95).

3.4. Case Studies of Renewable Energy Cities in Asia: Geothermal Energy

Geothermal energy is an abundant natural heat source that has been used as a source of electricity in over 24 countries and used for district heating in as many as 70 countries. It is usually cost effective, reliable, sustainable and environmentally friendly. For the past few decades many Asian countries have been planning various geothermal energy projects.

3.4.1. *Xianyang, China*

Located in Shaanxi Province, Xianyang is projected to be China's 'Geothermal Energy City' (PRLog, 2010; Verkís, n.d.). It has vast reserves of good quality underground thermal water. The Chinese government, in partnership with the government of Iceland, signed a Geothermal Energy Cooperation Agreement which set the planning conditions for Xianyang to become an ecological city promoting the development and utilization of geothermal energy.

The use of geothermal for district heating began in 2006 in Xianyang's university garden, in an area of 2.4 km^2 which is home to a number of schools, universities and their students. Numerous facilities that use geothermal energy have been established in Xianyang and the district implemented advanced technologies that use geothermal including geothermal wells networking, auto control and thermal water recharge. A development is being planned for 2012 in Fenghe which will have a number of housing projects and office buildings which will use geothermal heating for space heating and for domestic hot water. A geothermal energy association was established in Xianyang and is responsible for a number of demonstration projects showcasing comprehensive utilization of geothermal energy.

3.4.2. *Geothermal Energy in the Philippines*

As the world's second largest producer of geothermal energy, next only to the United States, the Philippines is currently carrying out various programs to further utilize this renewable energy source. The Philippines, located along the Ring of Fire, has a massive supply of geothermal heat. The country is home to a number of active volcanoes and hot springs (Figure 3.13). This massive reservoir of underground heat is now being tapped to supply clean energy to the country (Clough, 2008; Harden, 2008).

As early as 1967, the government had already enacted the Geothermal Law and a year later the first geothermal reserve was discovered in Tiwi Albay, where one of the main geothermal power plants in the country is currently located. Today, there are eight geothermal plants across the country in full operation, producing electricity equal

Figure 3.13 The Philippines Has a Huge Amount of Thermal Heat Reserves.

Source: Lee, 2010b.

to 28% of the total national electricity production. The country now has over 1,900 MW of installed geothermal capacity with government targets to double this production in the next decade to make Philippines a frontier in geothermal energy production.

3.5. Conclusion

Renewable energy sources have been proven to enable a number of economic and environmental benefits and now play an important role in sustainable development. Through the renewable energy examples presented, we have seen that a number of Renewable Energy Cities are now emerging and also that many more will be built in the coming years. Asia, which may have started a bit later than their western counterparts, is now more than keeping up and is putting in great efforts to promote the use of renewable energy. More Asian countries and cities are now dedicating themselves to the development, utilization and promotion of various renewable energies aspiring to become model cities for the future. They are hoping to inspire more cities to follow them in their quest for independence from fossil fuels and thus helping to demonstrate the new green urbanism of the future.

CHAPTER FOUR

The Bioregional Carbon Neutral City

4.1. Introduction

Carbon neutral means that city economies focus on their local bioregions to reduce energy and to use more and more local renewable sources of energy as well as offsetting the remaining carbon through innovative bioregional projects. In Asia there are a number of emerging case studies where carbon offsetting, sometimes through international UN mechanisms like the Clean Development Mechanism (CDM), are showing how new carbon neutral developments can be introduced into cities and regions. The Maldives has announced in 2009 that it aims to be carbon neutral by 2020. In Asia, the move to become carbon neutral is less seen and heard but when viewed in detail, small stories of hope emerge as people struggle to ride the next wave of innovation — green urbanism.

This chapter discusses the examples of:

- Azerbaijan: Zira Zero Island;
- China: Wanzhuang;
- India: Auroville; Tribal Development Center (Bhasha Center), Tejgadh, Baroda; Druk White Lotus School, Shey, Ladakh;
- Myanmar: Hlawaga Park;
- The Philippines: Puerto Princesa City, Palawan; La Mesa EcoPark, Quezon City;
- Singapore: Energy Policy;
- Thailand: Pig Farms and the Clean Development Mechanism; and
- Other Stories in Asia.

4.2. Zira Zero Island, Azerbaijan

Zira Zero Island is a project in Azerbaijan, aimed at enabling the island to be entirely independent of external resources including being carbon neutral. The island of Zira, located at the south of the main island Baku in the Caspian Sea, is 3.1 km long and 600 m wide. It is abundant in three resources: water, wind and sun (Averositi Holding, 2009).

A master plan put forward by BIG architects taps into these renewable energy sources, taking advantage of the island's natural landscape. The plan for the architectural landscape is for it to resemble the Seven Peaks of Azerbaijan and, according to Bjarke Ingels, the founding partner of BIG, it will create "an autonomous ecosystem where the flow of air, water, heat, and energy are channelled in almost natural ways" (Property Wire, 2009, n.p.n.). The developer, Averositi Holding, hopes that the master plan for the island will set a new standard for both luxury and sustainable living in Baku. Lars Ostenfeld Riemann, the Group Director of Building and Design at Ramboll, the engineering firm behind the development, maintains that "in a society literally built on oil this will serve as a showcase for a new way of thinking — sustainable planning" (Property Wire, 2009, n.p.n.).

The island will utilize many different technologies in order to be sustainable. Energy will be provided through offshore wind farms, located on the existing platforms and foundations of an offshore oil refinery. The heating and cooling of the buildings will be by heat pumps. Hot water will be provided through solar heat panels integrated into the architecture. The island will also incorporate photovoltaic solar panels into the building façades and rooftops to provide power for daytime functions. Transportation within the island will be provided by electric cars and taxis (however getting to and from the island requires a great deal of fossil fuel energy as it is only possible through the use of helicopters and boats). The island will utilize wastewater and stormwater for irrigation, with the solid parts of the wastewater composted for use as fertilizer. This process of recycling wastewater for irrigation and fertilizer will support the greening of the island while reducing the ecological footprint.

The architecture plan of Zira Zero is exemplary in many respects including its carbon neutral goal and emulation of the natural landscape of Azerbaijan. However, the transportation to the site and the building materials would produce significant CO_2 emissions due to the island's remote location. Thus, a truly carbon neutral project at Zira Zero would offset this carbon as is done in Masdar. Perhaps the planned greening of the island will already go some way towards this. The use of the term carbon neutral can be used to help draw together such ideas and focus the innovation.

4.3. Wanzhuang, China

A 30-year development framework is underway by Arup and Shanghai Industrial Investment Corporation (SIIC) for a new carbon neutral town extending from the existing City of Lang Fang, between Beijing and Tianjin in China. The new area would be called Wanzhuang, and encompasses an area of 80 km^2 with a population of 40,000 located in a series of villages. This area is predicted to increase to 400,000 people, with the city needing to expand without destroying the surrounding agricultural land. The plan replaces an unpopular proposal by the local Lang Fang authority, with the authority adopting the key principles of the new carbon neutral town (Arup, 2011; RIBA Journal, 2009).

The plan presented by SIIC and Arup is for an 'agricultural city' that protects agricultural land by developing compact cores of new development within the existing village footprints surrounded by 'ecological corridors'. These ecological corridors protect existing agricultural land and also provide additional land for new agricultural development and wildlife corridors. The new development will be up to six stories and will utilize an underground pneumatic system for waste disposal and renewable geothermal energy. A dedicated public transport loop is planned to connect the villages, with all residences to be within a 3 to 5 minute walk of either a bus stop or a bicycle path.

The benefits of the Wanzhuang plan goes further than providing food production close to the city. The ecological corridors will provide land for wildlife habitats and space for alternative transport such as walking and bicycling. The plan also encourages the use of farming that adopts ecologically sound production. It will preserve 65% of the existing agricultural land. Farmers will be given a choice between staying in their current village and relocating to a new urban area or to a rural farmers' collective. It is estimated that only 2% of waste produced by the villages will go to landfills.

4.4. Auroville, India

Founded in 1968 by Mirra Alfassa and designed by Architect Roger Anger, Auroville, meaning 'city of dawn', is a 'universal township' of 2,200 residents (2008) from 43 nationalities located in Southern India, in the Viluppuram district, Tamil Nadu (Auroville, 2010). The township was designed to accommodate up to 50,000 people as an experimental 'universal' town, accommodating diversity, alternative architecture, design and innovations along with fostering social, spiritual and environmental consciousness (Sobo and Hoberg, 2010).

The township and community are focused on sustainability and green living, with the tenets of living centered on unity of the mind, body and nature. The core planning principles are simplicity and harmony, with simplicity focusing on lifestyle choices and design, and harmony focusing on the use of environmentally sound technologies. The design focuses on incorporating and working with the local environment. The architecture of Auroville is highly bioregional, experimental and innovative, taking advantage of local building materials, climatic conditions (sunlight and wind) and topography. Furthermore, the architecture focuses on effective buildings that are able to respond to user needs with minimal footprints.

Auroville is home to several research institutes focused on modifying and integrating new and existing technologies and innovative processes to reduce energy and water usage. Of particular focus are solar technologies for energy production and for pumping water, cooking, street lighting, heating, and water management and recycling technologies. These technologies are all designed around being simple, affordable and efficient in energy use.

Solar energy is the largest renewable source of energy in Auroville. Some of the buildings run completely on custom-designed solar electricity produced by photovoltaic panels, with integrated inverter and battery storage systems. Currently, there are 400 houses running solely on solar electricity. The majority of the solar technology utilized is for pumping water, including in the wastewater systems, and for heating.

A large communal kitchen, the 'solar kitchen', is powered by a hybrid system that utilizes solar power and diesel fuel using the world's largest 'solar bowl'. The solar electricity system is a bowl composed of hundreds of mirrors that focus sunlight onto a heat receiver which is surrounded by coils filled with water. This water is turned to steam that is captured and then used for cooking. A diesel system is used when the solar electricity generation is too low to support the kitchen operations. The kitchen currently serves on average 1,000 lunches per day.

The Auroville township utilizes customized decentralized wastewater management and recycling technologies developed by the Center for Scientific Research (CSR). The wastewater system is composed of underground tanks for containment and pre-filtration and various above-ground systems for oxygenating. The institute has designed an efficient grey water recycling system that utilizes centrifugal and centripetal forces in acylindrical vortex system that only uses minimal space to filter and oxygenate the water. Effective microorganisms (EM) is an organic liquid composed of microbes used to quicken the decomposition of waste in wastewater treatment. Participation and community engagement compose a large part of the sustainability of the Auroville township. Furthermore, the township focuses on hiring local workers from the surrounding villages. These workers, often hired as laborers or service providers, are provided with training. This has increased the standard of living of the surrounding communities, with an estimated Rs. 900,000 ($18,000) flowing from Auroville into the surrounding villages every week.

Auroville does a lot of revegetation and rehabilitation in the surrounding bioregion though this is done for its own sake rather than for the purpose of neutralizing CO_2 emissions. As with Zira Zero Island, the full qualification of carbon neutral principles has not been done in Auroville but it has most of the elements in place.

4.5. Tribal Development Center (Bhasha Center), Tejgadh, Baroda, India

One of the most prominent problems in India is the lack of basic infrastructure in rural areas. Much of the focus of the authorities is in developing urban areas and concentrating development in urban cores. This focus also facilitates rural to urban migration. If government authorities adopted a more Gandhian outlook regarding the development of rural areas it may decrease some of the pressure of the migrating populations to urban regions. The Bhasha Center in Tejgadh, located about 50 km from the city of Baroda in Gujarat, India, provides an example of a rural community development project.

The Bhasha Center was established in 1996 as a Public Trust for the conservation and study of *Adivasi* (a term for the indigenous societies located in India) languages and culture by the Bhasha Research and Publication Center with the idea of establishing a leading institution for tribal studies (Adivasi Academy, 2006). The Bhasha Center is composed of a university, a research center, an activist center, a laboratory for experiments in development and a cultural expression forum, all designed to address all aspects of Adivasi culture and development. The Adivasi Academy was formed as part of the Center in 1999 and has been recognized as a Center of Excellence by the Ministry of Tribal Affairs, providing space for intellectual and creative expression, and importantly as a center provides hope for the future of Adivasi societies.

The Bhasha Center was designed by renowned Indian architect Karan Grover. One of the major criteria of Grover's design for the center was that it should reflect local character and relate to Adivasi culture. The whole center revolves around the idea of creating social awareness and development, including providing employment. It utilizes sustainable design solutions, particularly the use of local materials, enabling it to have a low carbon footprint. By providing employment, the center can help to lessen the migration of residents from this region to urban areas.

Cultural survival, through centers such as this, helps indigenous people around the world defend their land, their languages and their cultures. The Bhasha Center would find carbon-neutral status relatively easy to pursue and would be ideally suited to manage this kind of commitment.

4.6. Druk White Lotus School, Shey, Ladakh, India

The Leh Valley is in a remote mountainous area in Ladakh, India. The area is cut off from the rest of India for six months or more each year by snow. It is a high-altitude desert environment dependent on snow melt for water. The area is facing many challenges including water shortages and rapid urbanization with rural to urban migration causing issues of culture loss and skill shortages in the rural areas. In 1992 the Druk White Lotus School (also referred to as Druk Padma Karpo Institute) was initiated in the village of Shey, 13 km from Leh, the capital of Ladakh, under the guidance of spiritual leader His Holiness Gyalwang Drukpa

(Druk White Lotus School, 2012). The goal was to develop a school that could offer a high quality, modern education, a community center as well as be ecologically sustainable itself. The school opened in 2001 with a second branch in Khachhey in 2010. In 2011, the school had 650 students enrolled from the nursery through to secondary class 10, including many children from rural areas. About half of the students live on campus. The school aims to cater for 750 students when completed. The education provided by the school focuses on modern core subjects such as English, Maths, Social Studies, Information Technology, Physical Education and Creative Arts, along with subjects to ground the students in their own culture including Hindi and Bodhic (the local language).

The Druk White Lotus School was designed by Arup London using ecological principles: passive solar design, local materials and local labor. The buildings are designed to be operable year-round, to cope with the area's extreme temperature variations and to withstand the frequent earth tremors and heavy snowfall. The buildings are constructed out of granite with a mud core. This traditional material provides adequate insulation and fits with the context. The buildings feature indoor cottage gardens enabling students to grow food. To save water, the school utilizes recycled water and uses low-water technologies (or technologies that do not use any water), such as the Australian Ventilation Improved Pit latrines (VIP latrines) which not only eliminate fly and odor problems but also do not require water and produce humus which can be used as fertilizer. The VIP latrine utilizes a double chamber system with an integrated solar flue for composting.

The school is almost self-sufficient in terms of energy. The energy for electricity and for pumping the groundwater is provided through solar power. The buildings utilize passive solar heating with teaching spaces at a 30° southeast orientation which is optimal to capture the winter sun, and also feature fully-glazed solar façades. The façades store the sun's heat in high thermal mass walls. The residences are oriented south and utilize a technology called Trombe Walls which has an external wall constructed of a high thermal mass material and is faced with a double layer of glass. This enables heat to be stored in the walls and conducted inwards to the dormitories at night.

Water is stored in reservoirs uphill from the site and then gravity is used to feed the school with one reservoir providing drinking water and the other providing water for irrigation. These reservoirs enable snow melt to be stored and play a key aspect in the hygiene education program. The building structures utilize timber frames that are independent of the walls, steel connections and cross-bracings in order to resist the seismic loads of earthquakes.

The Druk White Lotus School won a World Architecture Award in 2002 for 'Best Green Building in the World', illustrating local initiatives to help mitigate climate change. The school also serves as a community center in the evenings and helps the community to be socially and culturally sustainable. The school also has an independently-accredited carbon-offsetting scheme, the Druk White Lotus Solar Energy Scheme, that people can use to offset their carbon emissions from traveling. Funds donated to this scheme are restricted to financing solar energy equipment (Druk White Lotus School, 2012).

4.7. Hlawaga Park, Myanmar

Myanmar, with a total area of 676,577 km^2, has a wide range of topography, temperatures and rainfall levels. The area contains many diverse ecosystems and many different elevations ranging from sea levels to high mountains. About 50% of the country's total land area is covered with biodiversity-rich forests. The government of Myanmar is now trying to conserve their biodiversity through different conventions, legislation, policies and plans. Conservation efforts are being demonstrated in Hlawaga Park using the carbon neutral concept.

Established in September, 1982, Hlawaga Park is located in Tankkyan Village, Mingaladon Township, Yangon division. The park includes a wildlife park of 818 acres (331 hectares), a mini zoo of 62 acres (25 hectares) and a buffer zone of 660 acres (267 hectares) (Trekthailand, n.d.). The objective of establishing the park was to provide an environmental education center, forest protection and to maintain a representative collection of Myanmar's indigenous wildlife, including species of mammals, reptiles and birds in an environment as near as possible to its natural condition. Other conservation efforts include effective protection and management of the degraded forest, plantations creating a buffer zone, the construction and maintenance of establishments in harmony with the natural surroundings, research studying vegetation, ornithological research on both resident and migratory birds and environmental activities to convey public awareness. These programs and activities contribute to restoring a once degraded forest into a natural forest environment along with providing opportunities for future research and public recreation. The next step is to seek carbon offsetting to help these goals.

4.8. Puerto Princesa City, Palawan, The Philippines

Puerto Princesa City, on the island of Palawan in the Philippines, is a bioregional area with a rich agglomeration of natural resources. The city is home to rainforest, mangroves and pristine sand beaches (Puerto Princesa City, 2010). Conservation programs by the local government have made the protection of these areas possible. There was a time, however, when none of these programs were even considered. In 1992, the city elected Edward Hagedorn as mayor. Hagedorn had a vision to see Puerto Princesa as a model city of sustainable development. He made the city one of the cleanest and greenest in the country through a program called Oplan Linis, which was a partnership between government agencies, non-governmental organizations (NGOs), private sector organizations and citizens. The program cleaned the streets, combated pollution and provided information and education. A hefty fine was imposed on littering and children were educated about the necessity not to litter (Local Governments for Sustainability (ICLEA), 2004).

When the mayor took office, one of his priorities was to provide decent homes to low-income residents that lived along the coastal areas of Puerto Princesa. The houses provided were designed to be liveable, affordable and energy-efficient, utilizing natural lighting and compact fluorescent lamps (CFL) instead of incandescent bulbs, together with enhanced natural ventilation, the energy demand of the buildings was reduced by 30–40%. A rainwater catchment facility was installed to reduce the demand for water pumping. The use of wood in the buildings for roofing, flooring, doors and staircases was prohibited in order to reduce the demand for logging. The anticipated annual savings from the use of the CFLs alone was predicted to be 21,414 kWh, equivalent to approximately US$35,610.

In 1999, Puerto Princesa joined the Cities for Climate Protection (CCP) campaign as one of the pilot cities in the Philippines. It pledged to reduce CO_2 emissions by 10% annually, the equivalent of 16,535 tonnes per year, against its 2010 forecast. As with the example of the affordable energy-efficient housing project discussed, the measures to reduce CO_2

emissions were practical and simple. These measures included reducing the hours of street light operations from 11.5 hours to 10.5 hours, resulting in a reduction of 80 tonnes of CO_2 annually; turning off lights and air conditioning in buildings during lunch break, reducing 4 tons of CO_2 per year; proving police officers with e-tricycles, reducing 44 tons of CO_2 per year; and a 'Feast of the Forest' annual tree-planting which would result in a savings of 7,800 tons of CO_2 per year.

In February 2010, the city installed a biodigester plant at the city's sanitary landfill. This waste-to-energy project is a partnership between the local government and the Philippine Bio-Sciences Company, a private firm providing renewable energy through biogas (World of Renewables, 2010). The facility collects waste from the city's sanitary landfill, wet markets, food establishments and slaughterhouses, converting it to electricity, producing up to 1 MW of power. The plant initially produced 350 kW of electricity on a 21-ton daily collection of organic waste and provides the power requirements of the e-jeeps and e-tricycles, completing the green loop (GMA News, 2010). E-jeeps can run up to 120 km on an 8-hour charge time while e-tricycles can travel up to 100 km on a full charge. A hydroelectric power plant will produce 6.8 MW of electricity. It is expected to reduce carbon emission by as much as 26,500 tons annually. Solar energy harnessing from photovoltaics is being considered to address the city's power shortage by a partnership with Optimal Power Solutions (OPS), a group specializing in distributed energy systems.

Resorts provide significant contributions to the local economy. SWITCH-Asia is a regional environment program that aims to promote the adoption of sustainable consumption and production among small- and medium-sized enterprises and consumer groups in Asia (SWITCH Asia Programme, 2011). SWITCH-Asia launched a project called 'Zero-carbon Resorts' in March 2010 which aims to showcase 'zero-carbon cottages' that utilize solar and biomass energy. The Cacao Pearl is one such resort positioning itself as 'carbon neutral' and uses passive cooling in the design of their cottages, along with wind turbines, solar panels and micro-hydro technology to produce renewable energy sources.

The measures used to reduce energy consumption in Puerto Princesa are very practical and implementable.

4.9. La Mesa EcoPark, Quezon City, The Philippines

La Mesa EcoPark is located in Quezon City, one of the most populous cities in the Philippines with 12 million residents. La Mesa EcoPark aims to rehabilitate, reforest, preserve and protect the La Mesa Watershed which is the primary source of drinking water for not only the city but also for the entire Metro Manila area. The La Mesa Watershed is a vital city feature not only as a source of drinking water, but also because of the forests within the watershed area. The forest functions as a lung for the entire metro area, providing it with clean air. This project is being realized through the joint effort of the ABS–CBN foundation (founder of ABS–CBN, the country's first and largest media network) and Metropolitan Waterworks and Sewerage System (MWSS). Together they formulated a partnership and created the strategy 'Bantay Kalikasan' (Nature Watch). Bantay Kalikasan actively involves the general public through a program of adopting a tree, a project or a hectare. This program has been a big success. Individuals become directly involved in tree-planting and private corporations fund the tree-planting activities. La Mesa EcoPark envisions a better environment for our children and future generations. It spreads environmental awareness through education and advocacy. It provides a healthy outdoor environment, a living classroom, a laboratory for environmental education, a center for biodiversity conservation and a popular destination for school field trips, family outings and picnics, proving a much healthier alternative to shopping malls (La Mesa EcoPark, 2012). It has not yet seen the potential to add to the carbon neutral status for the city — as with a number of other examples here. However, the potential to use this new funding source in the future will always be there.

4.10. Energy Policy, Singapore

Singapore is one of the leading cities in Asia in green urbanism. It has the potential of becoming a leader in sustainability due to its infrastructure, innovation and strong implementation of a well-planned urban environment. Although it currently does not have a detailed overall strategy of carbon neutral management yet in place, numbers of related government agencies are involved in advocating and initiating measures of carbon reduction. In 2009, the Singapore Government committed to reduce its carbon footprint by 2020. Various developments and innovations are arising to enable it to do this. Of particular interest are the carbon neutral developments, such as the 11 Tampines Concourse developed by City Development Limited (CDL), and the headquarters of Alila Hotels and Resort.

In November 2010, Singapore submitted its second National Communication on Climate Change report to the United Nations Framework Convention on Climate Change (UNFCCC) Secretariat, updating the first report submitted in 2000. The report details the strategies for managing sustainable growth and climate change in Singapore, reiterating the constraints in Singapore. These constraints include being a small, densely populated and urban city-state that is energy-poor, at an alternative energy disadvantage and has an export-oriented economy. The report highlights the key mitigation measures such as adopting less carbon-intensive fuels, increasing energy efficiency across households, industries, buildings and in the transport sectors (driven by the Energy Efficiency Programme Office) and investing in research and development for clean energy production such as solar energy.

In May 2009, the Economic Strategies Committee (ESC) was formed to "develop strategies for Singapore to build capabilities and maximize opportunities as a global city in a new world environment, so as to achieve sustained and inclusive growth" (Tay, 2010). The ESC is composed of members from the government sectors, the labor movement, the private sector and from academia. The ESC developed seven key strategies, one of which is for Singapore to become a Smart Energy Economy by February 2010. This strategy recommended that Singapore diversify its energy sources, enhance its infrastructure and system, especially investing in key energy infrastructure ahead of demand and to develop Jurong Island as an energy-optimized industrial cluster, to increase Singapore's energy efficiency, especially through promoting energy-efficient buildings, as well as in industries and in transport and residential use, to strengthen its 'green economy' through establishing energy as a key research and development priority, and finally to make sure that energy prices reflect their true total costs. These strategies aim to enable Singapore to build a smart energy economy that is sustainable (National Environment Agency of Singapore, 2010a; National University of Singapore, 2009–2010; Tay, 2011). In Chapter 6, an outline is given of Singapore's innovations in biophilic urbanism. This urban planting could enable significant offsetting towards carbon neutrality.

It is hoped that Singapore in the future will take the leadership and innovation in low-carbon technologies and biophilic planting and focus them into a carbon neutral framework.

4.11. Pig Farms and the Clean Development Mechanism, Thailand

Under the Kyoto Protocol, industrialized nations commit to reduce their greenhouse gas emissions to a certain percentage against the 1990 levels. Clean Development Mechanisms (CDMs) is one of the 'flexibility' mechanisms of the protocol from which industrialized nations can meet part of their cap by purchasing 'carbon credits' from CDM emission reduction projects in developing countries. Small pig farms in Thailand are beginning to see the benefits of participating in CDMs by pooling together CO_2 emission reductions and selling them as one unit in the carbon market. One such farm is the Moo 10 Village in the district of Chon Buri. As part of the Thailand AEP Livestock Waste Management Project, a community benefit plan is being established to improve the local living conditions. The quality of life in this coastal province of Thailand is very poor. There is only very basic infrastructure and very limited social services. There is no access to clean drinking water and no lighting on the street. Children cannot afford education and the decline of sugarcane production due to deteriorating soil quality has further compounded the poverty. Through carbon-financed projects, the community will be provided with a facility to produce clean drinking water and street lighting will be installed. Poor students will be given scholarships and a system to process livestock waste will increase income for the villagers. This project is the first carbon-financed project in Thailand. The project will generate 6,250 kWh of electricity per day using biogas harnessed from the swine manure from ten pig farms in the Chon Buri and Ratchaburi provinces. This will reduce the annual CO_2 emissions by 58,000 tons, with the reduction being sold as carbon credits to the Community Development Carbon Fund (CDCF), administered by the World Bank (World Bank, 2008). The potential for such projects across Asia is only just beginning to be realized.

4.12. Other Stories in Asia

There are many other ecological conservation and community empowerment projects in India, Bangladesh and Indonesia that can contribute to the carbon neutral city by means of carbon offsetting, reforestation and stopping urban migration. The following is a small sample:

- The Biligiri Rangaswamy Temple Wildlife Sanctuary in India is conserving energy through enterprising non-timber forest products. Partnerships with NGOs provided livelihood to the Soligas, a tribal community inhabiting the sanctuary, through enterprises in a honey-processing unit, a food-processing unit and a herbal-medicine-processing unit (Living into the Wild, 2012).
- The Nishorgo Project in Bangladesh seeks to conserve forests through carbon offsets with the involvement of the local community. They preserve a core area of wetland and forest, maintain a formal collaboration between the government and the community, and help poor communities by providing livelihoods.
- Tarakan in Indonesia is an oil exploration center where rapid urbanization is causing destruction of mangroves. With a vision for a more sustainable development, Mayor Yusof Serang Kasim implemented policies and supported programs to stop urban growth that was destroying some of the natural mangrove sites. This could have produced carbon offsets.
- The Toraja Native Settlement in Indonesia is an example of sustainable living in a rural area. They use manual carriages for transportation, grow local foods, use local building materials, and regenerate forests by planting trees when they have used wood to build their house. The bioregion is where the carbon offsetting efforts of the city is channeled to. The benefit is mutual in that cities can continually look to a future where their resource needs are met and bioregions retain their biodiversity and land area.

4.13. Conclusion

The stories presented here from different cities in Asia each illustrate techniques, policies, strategies and innovations to reduce carbon emissions focusing on carbon efficiencies, the use of renewable energy and on carbon offsets that enrich bioregional diversity. No matter how primitive or advanced the methods discussed are, they were bound by a common goal: to create a sustainable and resilient city that brings hope and enables the players to respond to the present wave of innovations. The notion of a 'Carbon Neutral City' was not formally used in many of these examples but could have been and there would have been advantages in the process, not least of which is the ability to accredit carbon offsets that will become more and more of an important source of funding. Nevertheless, the selection of stories are typical of many Asian settlements that are powerfully motivated to work more closely with their bioregions as they produce a more green type of urbanism for their future residents.

CHAPTER FIVE

The Distributed City

5.1. Introduction

The distributed city has its infrastructure more localized and scaled down to fit its local needs. Most Asian cities are already much more structured towards the notion of distributed infrastructure. They have usually grown in a way that combines local systems with regional systems for energy, water and waste. This is often because each step in development has been highly local in its ownership, governance and provision of economic and social services, though the regional advantages of big cities have often driven their continued growth and addition of more and more local parts. Now as green technology becomes available, it is feasible to refit these Asian cities with more effective energy, water and waste systems that build on the local structure of the city. This section discusses the efforts of four cities to implement distributed systems: Singapore, Tianjin, Jakarta and Dhaka.

5.2. Marina Bay, Singapore

Marina Bay is a new development in Singapore based on green urbanism principles. Although Singapore is highly centralized in terms of governance, many of its service provision elements are decentralized following a distributed city and cycle approach (Newman, 2011). Marina Bay, although celebrated as a tourist attraction, is not well known for its green features. Starting in the 1970s on reclaimed land, Marina Bay was envisioned to be a vibrant place where people will live, work and play (Figures 5.1 to 5.3). The planning of the area coordinates the land uses to ensure a balance of residential, office, education, medical and entertainment uses, and requires that all buildings meet the

Figure 5.1 Marina Bay with the Barrage, the Wet Biome, the Dry Biome and the Supertrees Visible.

Source: Peter Newman.

Figure 5.2 Marina Bay.

Source: Peter Newman.

Figure 5.3 The Marina Bay Area.

Source: Peter Newman.

Green Mark certification ensuring that they significantly reduce their consumption of resources. The Marina Bay utilizes many innovations to ensure a more distributed city approach to service provision, including utility provision through shared underground pipes, seawater air conditioning and the provision of many sustainable and interesting public buildings and spaces. The utility services at Marina Bay such as electricity, telephone, internet communications, water, cooling systems, rubbish collection, natural gas and sewage are all supplied through a common underground service tunnel. This tunnel enables easy maintenance and frees up space on the ground. Seawater Air Conditioning (SWAC) uses cold deep seawater and thermal induction and exchange to cool the water and air.

One of the most innovative elements of Marina Bay is the Marine Barrage Reservoir System. The Marina Bay reservoir, completed in 2008, is a freshwater reservoir formed by barraging the mouth of the Singapore River. The 350 m wide barrage supplies 10% of Singapore's freshwater to the surrounding regions, provides flood control and serves as a public attraction. The reservoir provides 55–60% of the water needed in the adjacent Central Business District (CBD) area and has helped to alleviate flooding in the low-lying areas of Chinatown, Boat Quay, Jalan Besar and Geylang. Water is treated locally at the barrage. The reservoir has also proved to be a popular attraction, primarily for recreational water activities. This is discussed further in Chapter 8.

Figure 5.4 The Supertrees at Marina Bay, Singapore.

Source: Gardens by the Bay, 2011.

To enhance the image of Singapore as a 'Garden City', three waterfront gardens have been built at Marina Bay (Gardens by the Bay, 2011). The 'Gardens by the Bay' features natural cooling through supertree technology, a dry biome and a wet biome. Supertrees are located along the edge of the reservoir, providing weather protection for entertainment and educational spaces. Supertrees are a natural cooling system that is provided by structures that incorporate vertical greenery (Figures 5.4 and 5.5). The supertrees grow ferns, vines, orchids and bromeliads, amongst other plants. The structures are fitted with environmental technologies mimicking ecological functions of trees, such as photovoltaic cells for lighting, rainwater collection for irrigation and fountains, and natural air flow. The dry biome is a large 1.2 hectare dome that will feature a permanent display of Mediterranean and semi-arid tropical plants and provide a space for flower shows (Figure 5.6). The wet biome is a small 0.8 hectare dome designed to replicate a cool and moist Cloud Forest environment typically found high in the tropical mountain regions (Figure 5.6). The wet biome will feature a 'Cloud Mountain' designed to enable visitors to experience a mountain environment.

The Marina Bay development has shown a comprehensive approach to planning supported by appropriate technologies. The area has been developed incrementally utilizing the innovative technologies available.

Figure 5.5 The Supertrees at Marina Bay, Singapore (close up).

Source: Peter Newman.

5.3. Sino-Singapore Tianjin Eco-City, China

The Sino-Singapore Tianjin Eco-city (Tianjin) is a collaboration between the Chinese and Singapore governments for a new planned sustainable city (Sino-Singapore Tianjin Eco-City, n.d., Figure 5.7). The city is located in an important national strategic area, the Tianjin Binhai New Area, in the districts of Tanggu and Hangu, and adjacent to the Tianjin Economic Development Area (TEDA), Tianjin Port, Coastal Leisure Zone, and approximately 150 km from Beijing. The site is well connected

Figure 5.6 The Wet Biome and the Dry Biome.

Source: Peter Nemann.

to the surrounding regions by good transportation networks and has good solar potential with solar radiation levels ranging between 80 to 240 W/m^2 and an average sunshine rate of 64.7%. When complete, the city's total area will be 31 km^2; it will have a population of 350,000 people and should be completed in the early-to-mid 2020s. In March, 2012, 60 families moved into the first part of the Eco-city.

The master plan for Tianjin is based around a distributed city concept of walkable eco-cells. The city is located in a fragile natural

Figure 5.7 View of Tianjin Eco-city.
Source: Ninbo University, 2008.

environment, containing saltwater marshes and a fragile natural ecology. The design for the city follows sustainability principles, using land, water and other resources efficiently, resource recycling, fostering independent innovation and protecting air quality. The framework for the city is based around a series of sustainability key performance indicators concerned with the ecological environment, social harmony and progress and with creating a vibrant and efficient economy. This framework is also focused on ensuring that the development is replicable, practicable and scalable. Some of the key green elements of Tianjin are outlined below.

Sino-Singapore Tianjin Eco-city is located in an area with freshwater scarcity, with low levels of rainfall. The local freshwater supplies will be unable to meet the projected demands of the city. Therefore, rainwater collection and efficient use of existing water supplies is imperative. The city will also utilize non-traditional freshwater sources such as saltwater desalination and water recycling. The goal is to use 120 litres of water per person per day, lower than in the greater Tianjin area (130.4 litres used per person per day), in Beijing (130 litres per person per day) and in Singapore (150 litres per person per day).

Rainwater collection will be done primarily through green roofs and will supply water for use primarily in municipal infrastructure. Green roofs will also enable rainwater to be filtered on-site through plants and soil before being collected. The green roofs will also provide natural cooling, reducing indoor temperatures and the heat island effect. The city will also utilize permeable pavements to increase rainwater infiltration into the groundwater.

The design of open spaces throughout the city is intended to alleviate land subsidence as well as reduce water logging and seawater intrusion through the use of low terrain and shallow vegetation in the wetlands along with public access. The wetlands act as natural purification systems.

Tianjin Eco-city seeks to minimize energy use and prioritize renewable and new energies, green buildings and sustainable transport technologies. The energy use key performance indicator for the city is 20% renewable energy provided through a variety of clean energy sources such as wind, solar and biomass. The city will have two Combined Heat and Power (CHP) plants which will supply electricity and heat. The renewable energy will mainly be utilized for the heating, cooling and domestic hot water of the buildings and also for street lighting.

The Eco-city will use street lighting powered by wind and solar. The street lights will be outfitted with a wind turbine, a solar panel and a battery enabling the lighting to operate on an independent system. The lighting also requires little maintenance costs after the initial construction and it has a long operational life span. Although the solar and wind lights have a higher capital cost, they have a longer life span, require less maintenance and use no electricity when compared to traditional street lamps, with the wind and solar lights having a ten-year maintenance cost of approximately 37% that of the traditional street lights and negate the need to purchase energy for street lighting. The city will require a total of 4,000 street lights, 2,400 of which will be the wind and solar street lights.

The Tianjin Eco-city will have a smart grid system to control the generation and distribution of energy. The smart grid system will also allow for an integrated monitoring system, an electricity information collection system, electric vehicle charging facilities and a communications and information network. Part of the system involves a fibre-optic communications network, which along with meeting electricity needs, will also enable communications through the system. To achieve this smart power grid, the telecommunications, internet and information services (such as high definition television) will be located in one network with four signals sent through the same optical fibre composite low voltage cable (OPLC). The smart grid will enable a two-way electricity system that can feed and monitor electricity to users and from users, allow for local renewables to enter into the grid system and for electric vehicle batteries to store electricity and provide electricity to the grid when necessary (called the V2G system — see Went *et al.*, 2008). This will be one of the first demonstrations of a major innovation in 21st century technology. When the project is completed, the communication network will cover the entire system from the power station to the user terminal, enabling intelligent electricity neighborhoods, intelligent home appliances and intelligence-powered business and will be a demonstration of how smart technology can enable a distributed energy system to work.

Sino-Singapore Tianjin Eco-city provides an example of a planned sustainable city utilizing elements of a distributed city model. These elements are exemplified by using land, water and other resources efficiently, resource recycling, fostering independent innovation and enabling localized infrastructure to be used in a smart, distributed way. It will be interesting to see how the city will develop and work once it is established. The smart grid system, in particular, could become a global model for 21st century infrastructure provision in cities.

5.4. Jakarta and its Surrounds, Indonesia

Jakarta is the capital of Indonesia and also the largest city there. Located on the northwest coast of Java with an area of 661 km^2 and a population of 9,580,000 people (in 2010), Jakarta is the country's economic, cultural and political center. It is the most populous city in Indonesia and Southeast Asia, and is the tenth largest city in the world. The urban area, Jabodetabek (Jakarta and surroundings of Bogor, Depok Tangerang and Bekasi), is the third largest urban area in the world with 28.0 million people in 2010, after Tokyo and Chongqing.

Social commentator Daddio (2009) maintains that the "[t]he city is increasingly known for its explosive population growth, seemingly endless traffic jams, polluted air and water, disappearing open spaces, urban poverty, and its vulnerability to frequent floods". Furthermore, "25% of the capital city's population lives in *kampungs*, slum neighbourhoods, spread around the city" and "another 5% of population lives in illegal settlements near public infrastructure or along riverbanks" (Figure 5.8). To help combat some of the city's problems, a number of sustainable innovations are being explored, such as distributed water systems and ways to develop sustainable and community-orientated residential areas and combat environmental and social issues.

Figure 5.8 A River-based Squatter Settlement in Indonesia.

Source: Peter Newman.

Jakarta has one water treatment plant located in Setiabudi, South Jakarta. The plant, established in the early 1990s, is only able to treat up to 3% of the city's wastewater. Furthermore, experts estimate that around 25–30% of the river water that goes into Jakarta's treatment plant does not meet official health and quality standards for untreated water (Cochrane, 2010). Potable water companies serve around 30% of Jakarta residents, with most residents dependent on groundwater taken from wells. The quality of this groundwater, however, is decreasing due to environmental degradation. Due to this, and the fact that the northern part of the city is sinking, city administrators have begun to phase out the use of the groundwater through prohibitive tariffs. The Jatiluhur Reservoir in West Java is literally Jakarta's lifeline, supplying up to 60% of the city's potable water needs. However, there are problems with quality and flow. The water levels at the Jatiluhur Dam are dangerously low. In 2010, the water level at the Jatiluhur Reservoir was lower than normal and as a result, freshwater provision was only 65% of the normal amount. These issues raise serious questions as to how residents in Jakarta will access freshwater.

Sewage and wastewater are also a large problem in Jakarta. Only 3% of Jakarta's area is connected to a sewage system. The World Bank costed a centralized sewage treatment plant, finding that the plant would be technically and politically impossible due to the disruption to the city (Diani, 2009). However, to combat some of the water and sewage problems, the city government has proposed a new sewage tunnel system that would channel household wastewater to a water recycling plant. The plant would turn the liquid-waste from septic tanks, kitchens and bathrooms into clean water able to be discharged into the nearby rivers and dams. The plan is to be completed by 2020, with the first phase of the project serving 700,000 people (10% of the total users) in a central zone running from Setiabudi (Central Jakarta) to a plant near Pluit Dam (North Jakarta). The second phase would provide wastewater services to an eastern zone, running from Gunung Sahari (Central Jakarta) to Sunter (North Jakarta). This would be followed by a third phase serving a western zone running from Palmerah to Kebon Jeruk (West Jakarta). After this phase the capacity of the sewage system would be 25% (Setiawati, 2010). This model of sewage treatment remains in the centralized mode and raises the question as to whether a distributed system would be better.

Water provision and water quality due to lack of water treatment has long been a concern in Jakarta. Proper water treatment in Jakarta has mostly been in the form of smaller scale facilities located in wealthy neighborhoods and provided privately. This is by treating groundwater and recycling wastewater. This small-scale approach could be an ingenious way to solve water supply and treatment problems, especially in Jakarta, as long as a cheap and effective way of establishing these localized systems can be demonstrated. Some demonstrations of low-cost systems have been provided through the World Bank (World Bank, 2012) and in other Indonesian cities (Newman and Kenworthy, 1999).

Lippo Village is a 3,600 hectares planned sustainable and community-orientated village of 10,000 families located in the suburbs of Jakarta. Lippo Village provides an example of water conservation and treatment within the Jakarta region (Figure 5.9). Within the village, all rainwater runoff is captured, contained and directed to water ponds for storage. These reserves help to keep the underground aquifer replenished. All of the buildings and houses in Lippo Village have piped freshwater and are attached to a central sewage system treatment plant which then recycles the water and diverts it to one of six water ponds located around the community to be used for irrigation (Cochrane,

Figure 5.9 Aerial View of Lippo Village Illustrating the Abundance of Green Environments.

Source: Lippo Village, 2012.

2010). Wahyudi Hadinata, the General Manager of Lippo Village's Town Management Division (TMD), maintains that the concept of using the waterponds is a simple but new idea for the government. The water storage and treatment facilities implemented at Lippo Village are very simple concepts; however, they need to be implemented through proper planning. To assist with this in a wider city context, Wahyudi suggests a regulation that all buildings (or at least all new buildings) collect, treat and use rainwater as well as recycle sewage water. This would be cheaper than using commercial water.

In addition to distributed water treatment, Lippo Village offers a practical example of a sustainable and independent township with a green environment, people-orientated infrastructure and flood-free setting. The village offers safe walkways for pedestrians, bicycle routes, open green spaces and outdoor cafes. A shuttlebus transports people around Lippo and also to and from Jakarta.

In 2008, the Lippo Village's TMD introduced a sustainable smart home. The home produces electricity through photovoltaic panels installed on the roof, uses geothermal energy to heat water and power the air conditioning, contains a small-scale water filtration and wastewater treatment facility on-site and minimizes the use of energy through natural ventilation and lighting.

The example of Lippo Village is clearly for the wealthy. It is what Newman *et al.* (2009) call an 'eco-enclave'. However, the village does illustrate what can be done technically in Jakarta. To develop this into something more extensive in Jakarta will require mass demonstrations to reduce cost.

Jakarta, like most Indonesian cities, has a lot of informal settlements. There have been many studies of how to make the most out of these places as they have strong social capital and many ingenious ways of managing water, energy and waste (Newman, 2009). The Kampung Code River is an informal settlement built on government-owned land along the banks of the Code River in Yogyakarta, Indonesia. Previously the site was a refuse dump. The informal settlement is comprised of 30 to 40 families whose members work in the nearby city market. Many of the squatter dwellings were constructed of plastic-covered cardboard. In 1983, the government wanted to demolish the informal settlement; however, community leader Willi Prasetya and a Catholic priest and self-taught architect, Yusuf B. Mangunwijaya, persuaded them to upgrade the site instead.

There were a number of concerns about upgrading the site. The area was subject to seasonal flooding and was very narrow and steep, separated from the river by a stone-retaining wall. The aims of the project focused on improving the quality of the river and the retaining wall, introducing services such as sewage to the informal settlement and on educating the residents about the importance of a clean and healthy environment (ArchNet, n.d.).

The first steps of the project were to rehabilitate the river and to reinforce the retaining wall against flooding. To do this, a better environment was created by planting the riverbanks with potted tropical plants. The next step of the project was to help formalize the settlement by building a community center, providing a place where people could gather to discuss issues and a place where school children could have a space to study. The community center was built from bamboo with mat flooring, with the exterior façades painted by local residents and volunteer art students.

The next step in the project was the introduction of electricity and a sewage system. This part of the program was fully supported by the government, with the government building communal toilets near the living areas and far from the river. Previously, the water from the washing and toilet areas would flow into the river. To avoid this, the

Figure 5.10 Current Code River Settlements.

Source: Proconservation, 2010.

government built a common septic tank and fitted new toilets with plumbing and vents for air circulation. The sewage from these toilets was then collected and taken to the city sewage. The third and final program was to provide education to the local residents about the necessity of a clean and healthy environment.

Kampung Code River now has a better environment (Figure 5.10). Although the scale of this project is small, the achievements made within the given constraints are immense and humane, making it a compelling model for other cities with similar problems.

Jakarta can learn much from these case studies that have successfully applied small-scale improvements. Jakarta consists essentially of many villages joined together, providing it with a structure in which small-scale projects focused on the community would be able to add up to significant changes.

5.5. Dhaka, Bangladesh

Dhaka, the capital of Bangladesh, suffers from many problems including transportation issues, pollution, freshwater shortages, inadequate sewage management and a lack of green spaces. However, there are many positive signs indicating that Dhaka is overcoming some of these issues. Some of these innovations include the Grameen Housing Program, efforts to make some of the factories (which employ large numbers of women) more sustainable, and the provision of green spaces.

Bangladesh is one of the top ten most populous countries in the world. It has a high poverty rate with about 63 million people (approximately 35.9% of the population) living below the poverty line (United Nations Development Programme, 2010). The country is prone to natural calamities such as flooding, river erosion and storms; every year, a percentage of people become landless. It is very difficult for the poor to build and maintain houses resistant to this climate. The Grameen Housing Project and Micro-credit Program are two programs trying to help combat this housing problem (Grameen Bank, 1984; 2011). The Grameen Housing Project was developed for the disadvantaged and landless residents of Bangladesh following the successful Grameen Bank Micro-credit Program.

The Grameen Bank, a non-governmental organization, first started a micro-credit program to enable people to start small enterprises. The micro-credit system was widely accepted and has since been utilized around the world (Microfinance Information Exchange, 2012). The majority of the borrowers in these programs have been women. By providing support and not charity, the loans help to restore people's confidence, health and well-being. The Grameen Bank enforces strict guidelines that all borrowers must follow, such as sending children to school, growing vegetables and looking after the health of children.

Following the success of the micro-loan program, the Grameen Bank developed a housing loan program for landless people. The scheme provides housing loans of between US$250 and US$600 at a 5% interest rate for previous Grameen Bank members that have lost homes. The repayment rate is near 100% and is done through a weekly installment of Taka 20 (US$ 0.30) for the first year, with special permission required if the borrower wants to pay more.

The loans require that houses to be built must be flood-resistant with waterproof and durable roofs, built with locally available materials. The Grameen Bank provides 4 concrete columns, 18 corrugated iron sheets, a latrine base with syphon and cement liners for each house, allowing the borrower to choose the other construction materials according to the affordability and availability of materials. The houses are restricted to 20 m^2 in area and follow a rectangular form with four columns on the four edges. The houses are constructed by the borrower. Effected communities are also eligible for a loan to build deep tube wells supplying access to drinking water. The tube wells can be shared among houses setting up a model for distributed infrastructure in a Bangladeshi settlement. The project received the Aga Khan Award for architecture in 1989 (The Aga Khan Award for Architecture, n.d.).

The garment industry in Bangladesh is a major part of the economy, contributing significantly to the GDP; and Bangladesh is one of the largest apparel-exporting countries in the world. The garment industry is a major employer of women and has made significant contributions to improve the livelihood standard of the rural and urban poor (Haider, 2007; Kamal and Zunaid, n.d.; Rahman, 1999). Many of these factories are making efforts to contribute to sustainable development in Bangladesh, particularly through the use of renewable energy and through reducing electricity consumption.

'Section 7 Ltd', in the Chittagong export processing zone, installed solar panels along its roof, installed LED lighting and built the walls of the factory with an internal cavity allowing for cold water to be passed through, cooling the building. Large water storage tanks were constructed in the basement allowing for rainwater to be stored on-site. These measures have helped to lower the electricity requirements, particularly the energy used to cool the building. Although not completely self-sufficient, the factory is an example of an efficient system, independent of the centralized water supply and with minimal dependency on the centralized electricity grid.

Medlar Fashion in Ashulia, Dhaka, has implemented an innovative cooling system in the rooms where the garment workers perform their tasks. A metal screen covered by a layer of flowing water was installed on one wall with exhaust fans installed on the opposite. This system passes air entering the room through the water screen, cooling it down, and thus enabling the factory to maintain a cool temperature for the employees without air conditioning. Further to this, the electricity used by the factory is supplied by a gas-powered generator with a transformer connected to the local power grid, able to provide standby power if necessary. The innovations in the garment factories such as Medlar and Section 7 Ltd illustrate examples of small-scale, distributed technologies that are context-specific and low-cost.

The provision of green space in Dhaka is very important to the sustainability and liveability of the city. Dhanmondi Lake is one of the most successful public areas in Dhaka. This space, along with providing public space and access to water, performs an important role in flood mitigation. The public space area includes recreational and exercise space, along with more formal public spaces such as an amphitheater, the 'Rabindra Sarobor', a sports complex with open fields (including one dedicated to women), food and beverage facilities, as well as many walkways and bridges.

The area surrounding Dhanmondi Lake contains a number of shopping centers, boutique shops, restaurants, offices, private institutions, schools, colleges and universities, attracting people from all over the city. There are plans for the area to be upgraded and for some distributed city innovations to be implemented. A recently constructed Meghna residence on Dhanmondi Road 10A is a good example of biophilic housing, with a green roof and a roof pond along with extensive use of green landscaping. The residence has been oriented and designed to utilize passive cooling and proper sun shading, reducing the amount of electricity required to cool the building and slowing down stormwater. The Dhanmondi area is quite successful in terms of its social and economic activities. If future developments also include environmental and infrastructure innovations, the area could become a model for a distributed city in Bangladesh.

The City of Dhaka suffers from a lack of water supply, poor sanitation, air and water pollution, traffic congestion and a lack of public spaces. It can be difficult and expensive to solve these problems at a city scale. The city can be divided into smaller areas which can be dealt with as smaller self-sufficient ecosystems. The city has already implemented measures such as those small-scale programs discussed above. These small innovations can be linked together to facilitate larger solutions.

5.6. Conclusion

Marina Bay in Singapore, Tianjin Eco-city, Lippo Village in Jakarta and Code River in Yogyakarta and Dhaka in Bangladesh all provide examples of the distributed city, focusing on small-scale distributed infrastructure and localized solutions, all requiring local ownership and enhancing local pride. Many of these solutions now incorporate green technology looking at the effective management and use of water, waste and energy.

CHAPTER SIX

The Biophilic City

6.1. Introduction

Biophilia literally means 'love for our living systems'. It is a term popularized by Edward O. Wilson in his book '*Biophilia*' (1984), which he describes as an innate affinity that human beings have with nature. He stresses that humans co-evolved with nature and so need it in their daily lives. Tim Beatley has applied the idea to cities and has tried to find what he calls a "daily dose of nature". The biophilic city therefore brings landscaping into and onto buildings, walls and roads to bring nature into every element of the built environment (Beatley, 2011).

The importance of biophilic urbanism is that it provides the possibility for dense cities to be able to regenerate nature and create far more natural urban systems. This is a big issue as the global planning world has generally recognized the need for increased densities to prevent car-dependent urban sprawls with all its oil, climate, health and economic implications (OECD, 2012; Newman *et al.*, 2009). However the need for natural systems to be part of this policy has always been a question that threatens to undermine the value in more compact cities (Newman and Jennings, 2008). Is biophilic urbanism a way to facilitate green and attractive cities that are also far more efficient in resources?

The need for a radically new approach to bringing nature into cities has never been more obvious than in the high-rise towers of the emerging megalopolises of the world, especially in Asia, which are usually surrounded by little more than grass and concrete, where biodiversity loss continues apace (United Nations Environment Programme, 2012).

Singapore has bucked the Asian high-rise tower syndrome by its planning over many years, particularly its recent commitment to

Figure 6.1 Singapore, a City in a Garden.

Source: Peter Newman.

biophilic urbanism. It now appears to be a leader in this new approach to city building (Figure 6.1). This chapter reviews biophilic urbanism by applying what appears to be happening in Singapore and then looks at some other case studies of Asian biophilic cities.

6.2. Singapore[1]

Since the early days of independence in 1965, Singapore has worked under a vision of a 'Garden City'. Many campaigns and slogans have been used to educate the public about the importance of keeping their city and the environment clean and green. In 1963 as part of the 'Clean and Green' campaign, the then Prime Minister Lee Kuan Yew initiated a tree-planting campaign with the aim of improving the air

[1] This story of Singapore has been made into a film: Newman, P, T Beatley and L Blagg (2012) *Singapore: Biophilic City* [Film], Curtin University Sustainability Policy (CUSP) Institute and Sustainable Built Environment National Research Center (SBE), available at http://www.youtube.com/watch?v=XMWOu9xIM_k

quality and marketing the 'Garden City' vision. To demonstrate the significance of this need to be a Garden City model in Asia, Lee Kuan Yew made National Parks (NParks) a part of the Ministry of National Development, at the heart of the new city-state's political agenda. NParks has since been a major source of innovation for the biophilic city in Singapore (National Parks Board, 2012b).

As the economy has developed, the people of Singapore's expectations for a better quality living environment with aesthetically-pleasing recreational spaces have increased. In addition to this, the increasing awareness of the need for environmental protection has resulted in a rethinking of the 'Garden City' model. This has resulted in the 'Singapore Green Plan 2012, beyond clean and green towards environmental sustainability', which shifts the vision from a 'Garden City' to a 'City in a Garden' (Figure 6.2) (Ministry of the Environment and Water Resources, 2002).

Figure 6.2 From 'Garden City' to 'City in a Garden'.

Source: Singapore Government.

At the 2012 World Cities Summit, the Prime Minister Lee Hsien Loong explained what this concept means,"In the next phase, our aim is to build a 'City in a Garden', to bring green spaces and biodiversity to our doorsteps." This seems to be the special feature that goes beyond landscape architecture and planning for parks and reserves: it means that buildings, roads and concrete are all potential sites for landscaping as well as the usual need to set aside green spaces between buildings.

The Singapore examples of biophilic urbanism set out below show that Singapore is doing both aspects of the biophilic city: creating more natural systems between buildings and also on the façades of buildings. These will be illustrated to provide a better understanding of how some of these biophilic concepts have been attempted in Singapore and are setting a global best practice.

6.2.1. *Biophilic Urbanism between Buildings*

6.2.1.1. *Regional Plans*

The Singapore Green Plan 2012 (SGP 2012), launched in 2002, is a government blueprint setting out Singapore's vision to be environmentally sustainable (Figure 6.3). One of its primary components is to conserve nature by replacing natural areas wherever development has disturbed it, providing information on indigenous flora and fauna through biodiversity surveys and creating new parks and park connectors (an island-wide network of linear parks and walking trails that links major parks and other areas; discussed below). Furthermore, the city has set up a National Biodiversity Reference Centre and in conjunction with the United Nation's Convention on Biological Diversity has launched the Singapore Biodiversity Index which is being used to measure the extent of biodiversity in cities across the world (United Nations Environment Programme, 2012).

The rationale for Singapore's commitment to greater regional natural systems is improving biodiversity, reducing the urban heat island effect, improving the thermal comfort of the outdoors, water management through reduced stormwater surges, and helping to reduce energy consumption in buildings (Yok *et al.*, 2009).

Figure 6.3 The Singapore Green Plan 2012.

Source: Ministry of the Environment and Water Resources, 2002.

6.2.1.2. *Streetscape Plans*

Creating a continuous tree canopy above all major roads is the backbone of the 'City in a Garden' vision and the Streetscape Greenery Master Plan (SGMP) is concerned with extending the local identity of an area, its sense of place in the physical landscape, to the stretch of road that spans the area, with the aim of creating a 'seamless green mantle' throughout the island. There are five distinctive landscape treatments based on ecosystem types: Parkway Treatment, Gateway Treatment, Coastal Treatment, Forest Treatment and Rural Treatment. The SGMP provided planning and design guidelines to achieve the variations in the character of Singapore roads and streets (Figure 6.4).

The need to conserve some of the more scenic and significant tree-lined roads in Singapore, particularly some of those with mature trees, is expressed through the Heritage Road and Heritage Trees Scheme, 2001. The current Heritage Roads are located in five areas: Arcadia Road, Lim

Figure 6.4 A Beautiful Continuous Street Tree Canopy in Singapore.

Source: Peter Newman.

Chu Kang Road, Mandai Road, Mount Pleasant Road and South Buona Vista Road (Figure 6.5). There is also an idea to extend Heritage Roads like Upper Thompson Road through Boulevards to give more impact.

6.2.1.3. *Park Connectors*

The Park Connector Network (PCN) is an island-wide network of linear parks that connects major green areas and destinations (particularly residential locations) around Singapore (National Parks, 2012a). PCNs use an integrated approach of 'greenery, conservation, education and recreation', aiming for a complete network of over 300 km of green connectors throughout the island by 2015. The plan is to make it possible to walk or bicycle around Singapore traveling through the parks (Figure 6.6). As of 2012, 200 km of the PCN had been built (Ministry of National Development, 2012) with a massive new north-south connector planned along the old regional rail alignment. This will link

Figure 6.5 A Heritage Road.
Source: Peter Newman.

Figure 6.6 Singapore's Park Connector Network.
Source: Peter Newman.

Figure 6.7 Singapore Botanic Gardens.

Source: Peter Newman.

the Central Business District (CBD) to the main green spaces across the island-city.

The PCN optimizes the use of underutilized land such as drainage reserves, foreshore and road reserves by turning them into canopy-covered green corridors for recreation and linkage for pedestrians and cyclists. One such example of these connectors is the West Loop Park Connector, which connects eight smaller parks. The Park Connectors offer a diversity of recreational activities and significant flora and fauna conservation (e.g. 550 species of butterflies and birds have been found there).

The Singapore Botanic Garden (established by Raffles at the beginning of the colony) is one of the pioneering institutions designed to link parks with people and to conserve indigenous and local biophilic knowledge (Figure 6.7).

6.2.1.4. *Horticulture Park's Biophilic Research and Development*

Horticulture Park, located in Southern Ridge Park, is the first gardening and lifestyle hub in Southeast Asia and the only garden-themed park

Figure 6.8 HortPark Research and Development Facilities on Green Walls.

Source: Tim Beatley.

in Singapore. The park preserves nature, maintaining biodiversity and attracting people to engage with nature through recreational, educational, research, commercial and natural developments.

Horticulture Park (HortPark) has been set up to demonstrate and experiment with green walls and green roofs to facilitate all aspects of biophilic urbanism. The various examples of green walls have become the basis for planners, designers and builders to experiment with how they have demonstrated biophilic urbanism for Singapore (Figure 6.8).

Green walls and green roofs, in particular, are being actively implemented through the city with NParks undertaking much research in the implementation of these within a Singapore context since 2003 (Yok and Sia, 2008).

6.2.1.5. *The ABC of Water Management*

The Public Utilities Board (PUB) looks after stormwater and has built concrete canals across the city for this important flood management function. They are now going biophilic by integrating water bodies and canals with parks and green spaces. The ABC (Active, Beautiful, Clean) Master Plan aims to enable water treatment as soon as the water touches the ground, enabling Singapore's water catchments, waterways, wetlands and the created bio-swales to have good water quality, be aesthetically beautiful, attract people and enhance the community. They also improve biodiversity and are an important biophilic feature.

Part of the ABC Master Plan is the improvement of Bishan Park, a joint project with Herbert Dreisetel's group. It is the first bio-engineering project to convert a concrete canal back into a river in Singapore and illustrates how the biophilic city can be achieved through water-sensitive urban design (Mouritz, 1987). The combination of plants, landscaping and innovative engineering, in a tropical climate, is a symbol for Singapore and any Asian city that tropical design can be biophilic.

The naturalization of the concrete canals (Figure 6.9) is done by demolishing the boundaries between the canals and the ground, and then using scattered concrete as part of the landscaping. Trees are replanted to filter the rainwater before allowing it to flow into river

Figure 6.9 Concrete from an Old Canal is Recycled into the New Kallang River, Singapore.

Source: Peter Newman.

banks. These naturalized waterways provide a water playground for families and naturally slow the water flow down. Emergency lights and sirens attached to a smart control system along the catchment enable clear directions to be given to the public whenever storms threaten to flood the waterway. Significant increases in biodiversity have been measured and the park receives 3 million local visitors in a year.

6.2.1.6. *Community Gardens*

The program, Community in Bloom, was launched in 2005 to help foster residents, workers and students to contribute to gardening in their community. Created by cooperation between Town Councils, the Housing Development Board, People's Association, the National Library Board, non-governmental organizations and the private sector, Singapore's gardening network comprises community gardening groups in public and private housing estates, educational institutions, organizations, charities, places of worship, community clubs, corporations and hotels.

The program is largely a response to the demand from community groups to have a closer daily connection to nature. The NParks staff sets up the groups, identifies a champion who can lead the garden group and together they seek the funds and volunteers to establish the garden. There are now 480 sites across Singapore. Increasingly, the groups are working on food production, often on top of buildings.

6.2.1.7. *Gardens by the Bay*

A new and iconic garden is the development around the Marina Bay Sands — the Gardens by the Bay — which features some extraordinary natural systems built with S$1 billion to regenerate a reclaimed foreshore. The Supertrees and display areas are all designed as educational features that show people how natural systems and cycles work. As a symbol of how nature can be built into a city, they are dramatic (Figure 6.10).

6.2.1.8. *How Has It All Worked Out?*

The greening of Singapore between buildings can be seen overall by comparing two satellite photos from 1986 and 2007, showing that the

Figure 6.10 The Supertrees at Gardens by the Bay, Singapore.
Source: Peter Newman.

green canopy cover has increased 20% despite a 70% increase in population. As with the community gardens which are moving onto building roofs, there is an increasing sense that the next major phase of biophilic urbanism in Singapore will be on the façades of buildings: green roofs, green walls and green balconies.

6.2.2. *Biophilic Urbanism on Building Façades*

Singapore's biophilic urbanism in landscaping buildings is found in both the public and the private sectors. This process has been enhanced by the introduction of planning regulations that require greenery in (and on) buildings, the Skyrise Greenery Initiative that subsidizes biophilic urbanism, and the Building and Construction Authority's (BCA) Green Mark Scheme that evaluates the sustainability of new developments.

Planning Regulations

Singapore has a strong planning system with a regional structure plan that sets out the park system and creates the park connectors, and a statutory scheme for each local area with guidelines for how much greening of buildings must occur. In the central commercial and shopping area such as Orchard Road, there is now a requirement that new or refurbished buildings must replace the whole footprint or floor plate coverage of a building with green space; this must be done with green roofs, green walls or green balconies. In this way the whole city is designed to have access to nature, no matter how dense or busy the built environment has become.

Skyrise Greening Initiative

In order to assist building owners and developers in the early phase of adopting the technology of biophilic urbanism through innovations such as green roofs and green walls, the Singapore Government has established a Skyrise Greening Initiative Scheme (SGIS). NParks assesses any proposal and pay up to half the cost of adding these green initiatives. 55 buildings in the period of 2009–2011 were granted subsidies under this initiative. At the beginning, the cost of greening was around S\$150/m^2, but after two years this came down to around S\$100/m^2. Several of the case studies set out below on greening buildings have been given the SGIS subsidy.

The Green Mark Scheme

BCA launched the Green Mark Scheme in 2005, as part of an effort to be green on a building scale. The scheme is intended to direct the construction industry towards more environmentally-friendly developments and assesses the construction process, the structure, the materials used and the performance (indoor quality and the use of energy) of the building. The Green Mark Scheme encompasses various building scale projects from residential to non-residential buildings, building renovations, office interiors, new and existing parks, infrastructure, district level projects and overseas projects. The Green Mark Scheme is increasingly incorporating biophilic elements into its assessments.

6.2.3. *Case Studies of Greening Buildings*

a. *Six Battery Road — Existing Building Going Green*

Six Battery Road is a high-rise office building located in Raffles Place, overlooking the Singapore River. The developer, CapitaCommercial, wanted to enhance the performance of the building to be greener.

In order to accomplish this, the developer tried to apply an efficient chiller plant, redesigning the chiller plant room system to achieve 30% efficiency improvement and a potential improvement in energy efficiency of up to 25%, and a self-sustainable green wall. The creation of the iconic 184 m^2 green wall was the biggest green wall installed in an office building in Singapore (Figure 6.11). The wall utilizes a rainwater harvesting system fed to the wall through an automatic irrigation system. The green wall was designed by Patrick Blanc, whose walls in Paris, London, Madrid and Sydney have helped to establish this new biophilic design element. The building utilizes wind turbines to

Figure 6.11 Green Wall in Lobby Area.

Source: Six Battery Road, 2011.

produce energy to power the green wall's irrigation pumps and for lighting. The building has a system of sensors that monitors carbon monoxide levels in the car park. This system brings in fresh air when the carbon monoxide levels increase beyond a certain point. Furthermore, the car park utilizes natural daylight through the use of solar light tubes and provides special parking spots for hybrid and electric cars. A 'green room' that displays real-time monitoring of the energy and water usage and their savings is open to the public and provides an education corner to create awareness for both the tenants and visitors.

b. *Republic Polytechnic — Campus in the Park*

Republic Polytechnic is a development that demonstrates an integration of sustainability from the project's inception to completion. One aim of the development is to ensure a pleasant environment for present and future occupants to appreciate nature, in line with the government's intention to transform Singapore from a 'Garden City' into a 'City in a Garden'.

The Republic Polytechnic campus was intentionally located adjacent to the regional park, in a secondary forest, enabling nature to be integrated into the campus. This connection to the regional park into the campus is provided through jogging trails, greenery and a soft but defined boundary between the campus and the regional park.

Other factors at the Republic Polytechnic campus also contribute to the biophilic concept. A compact plan was proposed to maintain as much of the existing greenery as possible. In doing so, 44% of the site's surface area is made of greenery. In order to increase and replace the greenery lost, the cleared areas are replanted to integrate the campus with the natural surroundings. The campus also contains a rooftop garden and a vertical green wall on the multi-story car park composed of creepers that grow from vine trays. The development also utilizes recycled building materials, including the recycling of biochips for compost. A building management system monitors utility usage and ensures the optimization of energy consumption. The energy savings of the building through the course of its life cycle is around 175,000 kW per year and a photovoltaic energy savings of approximately 11,500 kW per year.

Figure 6.12 Exterior View of Newton Suite.

Source: Tim Beatley.

c. *Newton Suite*

Newton Suite is an example of a high-rise vertical green wall. The building incorporates a lush green environment through the provision of exterior green walls, communal skyrise gardens and vertical greenery that extends 36 floors up (Chiang and Tan, 2009). The design was envisioned by the architects (WOHA) to be an attempt to advance the idea of a 'tropical' high-rise residential model (Figure 6.12). The greenery becomes a building material. In doing so, the building has a green area 130% of its total site area. The development demonstrates the idea of having 'land' and 'green space in the sky' concept of living and was given a SGIS subsidy and is the winner of the 2008 Singapore Institute of Architects (SIA)–NParks Skyrise Greenery Award (Chiang and Tan, 2009).

d. *Changi Airport Terminal 3 — Architecture of the Landscape*

Changi Airport is the city's gateway and was seen as the chance to demonstrate biophilic innovation. Changi Airport Terminal 3 reflects Singapore's vision to be a 'City in a Garden' by demonstrating an interior architectural landscape environment through the combination of building technology, natural lighting and a living green wall. The terminal features a 15 m high and 300 m wide vertical green garden wall, featuring 20 species of climbing plants found in a tropical rainforest, grown from planter boxes mounted on a steel frame (Chiang and Tan, 2009; Figure 6.13). The wall helps to create living spaces without borders between building and garden, architecture and landscape. The green wall not only 'greens' the terminal but also adds a unique identity to the place (ARCPROSPECT International Foundation, 2011).

Figure 6.13 Images of the Green Wall at Changi Terminal 3.

Source: Peter Newman.

e. *Khoo Teck Puat Hospital*

Located in Yishun, Singapore, Khoo Teck Puat (KTP) Hospital is a new 550-bed general and acute facility that opened in June 2010. It is the world's first biophilic hospital built with these features planned from the beginning.

The origins of this brave experiment are in the much older Alexandra Hospital. As the CEO of KTP Hospital Liak Teng Lit explains, he was asked to take over the Alexandra Hospital but found the buildings and general environment very clinical, drab and boring. He decided to appoint a Chief Gardener, Rosalind Tan, an occupational therapist (OT) in the hospital with good community links, and gave her the task of landscaping the hospital using voluntary labor (including a retired Professor of Botany). After three months the place was looking up and the staff began volunteering their time. The landscaping included creating a medicinal garden and a fragrance garden, along with water features. Birds and butterflies began returning to the site and they set a goal of having 100 species of butterflies return to the site. After three years of landscaping, 102 butterfly species returned. As Liak Teng Lit says, it was a transformation from "this hospital is haunted" to "this hospital is a butterfly garden".

When the Singapore Government began to plan a new hospital they asked Liak Teng Lit to guide the process and bring some of this 'butterfly magic' to the new site. A young architect Jerry Ong from CPG was appointed and although he knew nothing about biophilic design he quickly learned.

The new KTP Hospital features a green roof that has been built to withstand the extra weight and grows vegetables, fruit trees (140 trees) and spices that are managed by a local community workforce (Figure 6.14). The garden's produce is sold in the Hospital Canteen to pay for any garden costs.

Jerry Ong responded to the design brief of creating a 'hospital in a garden' by building green walls, green balconies and multi-level gardens (including ponds with 92 fish species) such that every bed and every office is surrounded by plants with 'healing views' (Figure 6.15).

The evidence for improved healing is yet to be fully analysed but all the anecdotal evidence points out that people heal quicker and are happier in KTP. Blood pressure and heart rates go down as soon as people sit in the tranquil environment.

Figure 6.14 The Rooftop Garden at KTP Hospital, Singapore.
Source: Peter Newman.

Figure 6.15 Green Views at KTP Hospital, Singapore.
Source: Peter Newman.

Figure 6.16 KTP Hospital, Singapore, is Popular with School Groups.

Source: Peter Newman.

The hospital is so popular with the general public that they have regular tours by schools and community groups; the park-like spaces are used by people who come just to sit and enjoy the space as well as students who bring their laptops and study there (Figure 6.16).

So far the KTP Hospital has found that 32 species of butterflies and 24 species of birds have made their home in the green spaces. The green roofs and landscaped areas capture and reuse approximately 12% of the rainwater runoff. The energy consumption of the KTP is 30% less than comparable new hospitals, saving S$1 million a year. But the health productivity gains from the biophilic elements will greatly outweigh this.

f. *158 Cecil Street*

Architect Kelvin Kan from AgFacadesign was faced with a real problem — a 14-story commercial building in Singapore had failed to attract tenants as the large façade on the front, built to reduce sunlight, was

Figure 6.17 158 Cecil Street, inside the Façade.

Source: Peter Newman.

ugly and impractical. His solution was to create a green wall like a 'Hanging Garden' camouflaging existing structures and creating natural ventilation for new green spaces on suspended balconies linking offices to the new space (Figure 6.17).

The green wall at 158 Cecil Street consist of 13,000 plants in 350 m^2 and a further 70 m^2 of hanging plants on balconies. This represents 135% of the building floor plate (not including the hanging plants) (Landscape Forum Nature and Living, 2011). The resulting green columns, green walls and green balconies were a stunning success. Not only is the wall home to butterflies and birds, but it also is a work of art. Looking from the bottom upwards, the green wall looks like a cathedral (Figure 6.18). As soon as tenants began to see the wall being built, they started turning

Figure 6.18 The Green Wall inside 158 Cecil Street.

Source: Peter Newman.

the focus of their offices towards the green space being created. Values of commercial rental space rose sharply as 'cool' companies chose to locate in this new space. The owners of the building were very happy with the rejuvenation of their asset (Landscape Forum Nature and Living, 2011; Southeast Asia Building, 2011).

One important feature of the design is how every part of the wall is easily accessible and each plant can be replaced or tended to individually as they are each in separate pots with their own irrigation point. Thus, maintenance is easy.

g. *Hougang Primary School*

Biophilic urbanism needs to infiltrate into all parts of the built environment, including institutions like schools, to enable them to be models for the future. Hougang Primary School in Singapore is showing that biophilic innovation can be trialled by young children at a very low cost.

Science teacher Mohan Krishnamoorthy was fascinated by green walls and wondered if he could make one at his school. So he took himself to Paris and London to inspect the work of Patrick Blanc. What he found was a simple technique where a wooden base could have a felt cloth cover stapled on with slits for soil added and stapled around; plants could then be slotted into these pockets and with drip irrigation, the wall would start to grow.

After setting up a trial, he was convinced the students could do it, so on one day in July 2011 he started with his special science class to build a green wall. The results were a beautiful and functional addition to a school that already features a range of gardens for teaching purposes (Figures 6.19 and 6.20). The students are very proud of their green wall and the school receives many visitors who leave convinced they too could make one.

Figure 6.19 The Green Wall at Hougang School.

Source: Peter Newman.

Figure 6.20 The Gardens at Hougang School.

Source: Peter Newman.

The significance of this case study is that it shows how biophilic urbanism is not really difficult or expensive; it is something that can be done by anyone if they have the mindset.

6.2.4. *Conclusions to Singapore*

Singapore is a good example of biophilic urbanism, where the development of green areas and green buildings are being shown to regenerate the natural systems in the city. The importance of their model is that many other Asian cities are beginning to copy their approach and enable their dense urbanity to be expressed in a more natural way. The fact that Singapore has achieved this quite quickly is a tribute to their commitment to innovations in urban planning. They have demonstrated that planning regulations and planning strategies for biophilic urbanism can be delivered cost-effectively and with strong community support, that demonstrations can be quickly mainstreamed, that government incentives and R&D are all part of the mix for enabling innovative change, and that political leadership drives everything. The next stage in the development of biophilic urbanism will be for cities with very different climate conditions to apply the same principles, and for developed cities like Singapore to mainstream the process by evaluating and quantifying the benefits in terms of energy, water, health, aesthetic qualities, human appreciation and economics.

6.3. Other Asian Cities

6.3.1. *Introduction*

The wave of being biophilic, or amalgamating nature into developments, is not limited only to Singapore but is also present in many other Asian cities. This section discusses examples from other Asian cities, looking at:

- The Green Tram, Kagoshima, Japan;
- Urban farming, Mumbai, India;
- Cheonggyecheon River and Parks, Seoul, South Korea;
- The conceptual works of Michael Sorkin;
- Masdar city center, Abu Dhabi; and
- Asian Crossroads Over the Sea (ACROS) Fukuoka Prefectural International Hall, Japan.

6.3.2. *The Green Tram, Kagoshima, Japan*

Kagoshima City is Kyushu's southernmost major city and the capital of Kagoshima Prefecture in Japan. The city has implemented a new Green Tram using two retired electric tram cars. The environmental benefits of the trams are obvious: they enable fewer cars for transportation, decrease congestion and the electric trains are more efficient and create less air pollution than their automotive counterparts. In Kagoshima, a volcanic stone called *midori*, when crushed and used as a base for gardens and lawns, is incredibly nutrient-rich. The result is an almost fluorescent and always flourishing green growth. The authorities in Kagoshima realized that if *midori* (green) was placed around and under the tracks, it would not only improve the appearance of the tracks but also help to reduce the heat island effect caused by the existing acres of concrete and pavement. The greening of the tracks has resulted in a dramatic temperature drop. Presently, greenery has covered about 3 km of tracks and the city hopes to extend this to more than 9 km by 2012.

In Japan, mayors and other top leaders of the local jurisdictions are looking to introduce next-generation light rail transit (LRT) to their areas. Kagoshima has shown that this can be done along with biophilic qualities.

6.3.3. *Urban Farming, Mumbai*

There are around 800 million people engaged in urban agriculture worldwide that contribute to the food needs of urban residents. According to Ben Bareja, editor of cropsreview.com, by 2025 the number of cities with populations exceeding 10 million people is expected to increase to 29, making a minimum of 6,000 tons of food needed to be imported daily into each of these cities to sustain their populations. Furthermore, he stresses that low-income urban residents spend a substantial proportion of their income on food, with the poorest people in the poor nations spending the greater part of their income, between 50 to 70%, on food (Bareja, 2010). However, due to such urbanization, the agriculture lands or fields around cities are becoming part of the urban area as cities grow, usually becoming residential areas. The demand for agricultural products increases while the available land for agricultural production decreases. This causes the need to import more food and grains from other regions, which not only increases the ecological and carbon footprints of the area, but also increases inflation issues for the population buying it.

These problems are now being faced by the city of Mumbai on a major scale, with the current level of urbanization 2.1% higher than in 1991. The Mumbai Metropolitan Region has 16.4 million residents and is the largest urban agglomeration in India. The problems regarding available land for food production have been traced by government authorities and they are trying to find a solution to such a complex problem. Two examples of authorities who are leading the way forward by working on propagating urban farming in the city are the railway authorities in Mumbai and the work of a number of non-governmental organizations (NGOs) within the area, including the work of Dr R.T. Doshi, the Urban Farming program and a school farm project by the Organization for the Protection of Children's Rights (OPCR).

Railway authorities have come up with a 'Grow More Food' scheme recently under which railway land is licensed to railway employees belonging to certain groups for the growing of vegetables and crops. This program has been particularly successful along the western line, which runs to Virar, where there has been a constant increase in the growing of vegetables and herbs. The urban farming along the rail alignment in Mumbai is still in bits and pieces but the project has tremendous opportunity to expand further and to be implemented in other parts of the Mumbai area.

Dr R.T. Doshi is an intellectual residing in Mumbai, experimenting with developing urban-based farming methods that would enable people to grow food in very small spaces, such as terraces and balconies. The methods all aim to be low-cost and low-maintenance, requiring little equipment. Dr. Doshi is able to grow many different vegetables, pulses, fruits and cereals on the terrace of his bungalow located in one of the crowded areas of Mumbai, harvesting things such as mangos, guavas, bananas and sugarcane. The innovative practices pioneered by Dr. Doshi have now been adopted by several households throughout Mumbai and neighboring Pune. This project is similar to an urban ecology project in Bogor, Indonesia by Hari Harsono Amir — see Newman and Jennings (2008).

The Urban Farming program initiated by Adrienne Thadani and Karen Peters in Mumbai is a program designed to spread knowledge and encouragement about growing food and herbs in your own house. Thadani and Peters have created a terrace garden at the Bombay HUB (a social network forum) where they conduct courses that take people through the basics of urban farming.

OPCR in Mumbai has an Urban Farming program designed for street children. The program creates farms in city schools. Through these farms, OPCR provides the street children with technical food production training and also with income from selling the farm produce, such as flowers, fruits and vegetables, in the retail market. The program aims to provide economic support for street children, to beautify the city's landscape and provide locally-produced organic food to urban dwellers and to demonstrate the ecological soundness and socio-economic benefits of city farming.

Box 6.1 Permaculture in Indonesia

The introduction of green revolution technologies (i.e. chemical fertilizers and pesticides) to Indonesia in the 1970s under the first Five-Year Development Plan brought the country closer to self-sufficiency in food production, especially in rice production. However, in the 1980s, the production of rice dropped, primarily due to pests becoming resistant to pesticides, and the country began to experience widespread soil and groundwater pollution of soil from the use of pesticides and the loss of genetic diversity (primarily in crop species). To combat this, Indonesia imported rice, banned certain pesticides and introduced integrated pest management and reforestation programs. Unfortunately, these programs were largely unsuccessful. An alternative approach was sought by local residents. Permaculture provided such an alternative.

Permaculture is an intensive form of agriculture that utilizes a wide variety of plant and animal species and organic pestmanagement, and works to maintain soil vitality through composting and earthworms. Permaculture gardens are rich in biodiversity and mimic the natural stratification of species. A similar approach, known as *kebon talun*, is very familiar to the Sundanese people from West Java, where it is regarded as a transitional stage between shifting cultivation and permanent agriculture.

A rehabilitation project aimed at utilizing and combining permaculture and *talun*, run by Harry Harsono Amir, was started in the 1990s in Sukabumi, in West Java. The project aimed to restore soil productivity and the livelihood of local farmers. Sukabumi is a large regency province with poor soils and a low per capita income. The primary livelihood within the area had been largely centered on tea, coffee and rubber plantations; however, as many of these plantations became unviable, they were replaced with monoculture farms. Through the permaculture project, Amir provided farmers with local species of plants and animals, including local varieties of chicken, species of medicinal herbs, bamboo, softwoods and hardwoods grown at a suburban plot in Bogor (Figures 6.21 and 6.22). In one initiative, the project provided 452 local chickens of different varieties to 115 household.

(*Continued*)

Figure 6.21 Harry Harsono Amir in His Permaculture Garden.

Source: Peter Newman.

Figure 6.22 The Village of Sukabumi in West Java, Indonesia, where Rare Species Have Been Reintroduced into the Village Economy.

Source: Peter Newman.

(Continued)

(*Continued*)

The project has had varying degrees of success. Some of the participating farmers have been very successful, with some even establishing businesses, such as the production of compost. Unfortunately, 60% of the farmers involved failed to establish ongoing production. Many were extremely poor, and had to consume or sell their products before permanent production could be established. Others found the market for their products limited and failed to establish a source of income. Although the project was not completely successful, opportunities still exist to implement permaculture techniques and more sustainable methods of local food production. These include cooperatives to provide credit for farmers, eco-tourism potentials and small-scale home industries. The project demonstrated how urban areas can contribute to the local ecology and protect local biodiversity.

Sources: Amir, 2000; Newman and Jennings, 2000.

6.3.4. *Cheonggyecheon River and Parks, Seoul, South Korea*

As the City of Seoul in South Korea expanded, the Gangnam area (the southern part of Seoul) was developed into a prospering new downtown, while the old CBD located in the Gangbuk area (north of Seoul) went into decline. This CBD area needed to be revitalized in order to be more viable and liveable. One of the primary factors believed to have led to the degradation of the downtown area was the Cheonggye Elevated Highway and the loss of the Cheonggyecheon River.

In the early 2000s, a vision for the revitalization of the Gangbuk area was put forward. The vision was for the area to be revived as an eco-friendly downtown, an area that combined the historical, cultural, business, commercial and tourism needs of the city. The restoration of the Cheonggyecheon River became the pilot project symbolizing the revitalization of the downtown and Seoul's transformation into an eco-friendly city. This remarkable story is even more astonishing when it is considered that the man who was in charge of building the freeway was elected to be the Mayor with a commitment to remove the freeway — and is now President Lee, the leader of South Korea.

The demolition of the Cheonggye Elevated Highway began in 2005. All the scrap iron and steel from the demolished highway, along

Figure 6.23 The Cheonggyecheon River Restoration.

Source: Peter Newman.

with 95% of the concrete and asphalt, were recycled. As a pedestrian-centric public space, Cheonggyecheon was designed to be a human-and-environment friendly space with sidewalks along both banks of the river. Furthermore, to reduce car emissions, a new traffic plan was introduced to reduce vehicular convenience and improve the public transport (Figure 6.23).

Today, the restoration of Cheonggyecheon has reintroduced the ecology of the stream with 213 species of plants, fish, birds and other organisms. It has also improved the surrounding environment's air quality, water quality, the noise levels and the heat island effect. The data on the effect of project is provided in Table 6.1.

Lastly, the Cheonggyecheon project has also created a more liveable environment in the city and a magical place where people can escape from their everyday work and living.

Following the success of the Cheonggyecheon River restoration, and in recognition of the severity of climate change, the Seoul Metropolitan Government (SMG) formed a global environment team in 2005 to be in charge of works related to climate change. To put the Government's

Table 6.1 Impact of the Cheonggyecheon River Restoration on the Environment and the Ecology.

Environment	
Air Quality	NO_2: 69.7 to 46.0 ppb (2006) PM10: 74.0 to 48.0 $\mu g/m^3$(2011)
Water Quality	BOD: 100–250 = 1–2 ppm
Noise Level	Reduced from 73.8 to 62 dB(A) during the day (2010)
Heat Island Effect	Reduced by up to 5°C
Wind Corridor	50% increase in average wind speeds
Ecology	
Fishes	5 to 25 species
Birds	6 to 36 species
Insects	15 to 192 species

Source: Centre National De La Recherche Scientifique and United Nations Convention on Biological Diversity, 2010; Lee, 2006; United Nations Human Settlements Programme (UN-Habitat), 2012.

commitment to create an eco-friendly city into action, SMG initiated several projects expanding Seoul's parks and green spaces by transforming brown-field sites. This commitment will be demonstrated by the Seonyudo Park and Nanjido World Cup Eco Park transformation projects, both part of the Han River Renaissance Plan and Seoul Beautification Plan. Furthermore, the Republic of Korea developed a National Strategy and five-year plan for 'Low Carbon, Green Growth' in 2008.

Seonyudo, an island on the Han River, used to be the site of a sewerage treatment plant. Upon the plant's closure in 2001 and after two years of careful planning and restoration, Seonyudo was transformed into an eco-conscious park. Today, the park harmoniously combines industrial elements with new landscapes, preserving the former treatment plant's structures as part of its landscape features. The basins previously used for water treatment were transformed into homes for aquatic species and plants that naturally purify the water. What used to be an inaccessible and polluted island in the middle of the city and river has now been transformed into an attractive park destination for locals and visitors alike.

The Nanjido World Cup Eco Park is located on a former waste dump site along the Han River. The site is being converted into an eco-friendly tourist destination and park, which will feature culture, arts, leisure and sports. It is planned to be the Mecca of eco-friendly city beautification projects in Seoul, and will contain a 'Zero Energy House' that will serve as a site for public relations for environment protection and renewable energy, along with a 'Hydrogen Station' able to supply hydrogen fuels for the buses at the park. The idea is to transform the park into a world-wide theme park for environment protection, adding elements of joy and fun while protecting the environment.

6.3.5. *Conceptual Works on Biophilic Cities by Michael Sorkin*

This section describes some of the conceptual biophilic urban designs by Michael Sorkin. These concepts are typical of how high density is increasingly being seen as requiring a simultaneous landscaping of the whole built environment.

Penang Peaks is a concept for a mixed-use urban development. The area would provide housing, office space, commercial space, medical facilities, a convention center, a transit node, a concert hall and a space for schools, universities and extensive recreational facilities. The concept is a model of best environmental practices. The architecture not only appears green but will also be environmentally green by being fully self-sufficient in water and waste management. The focal element of the concept is a 20-hectare park that is linked to the Penang Hills via a land-bridge over the new Penang Outer Ring Road. The park is to be surrounded by residential buildings, all connected by Kudalari Drive, which would be the city's premier retail and commercial boulevard.

6.3.6. *Masdar City Center's Courtyard*

The more recent innovations in the concept of biophilia are not only to incorporate natural elements but also to mimic natural elements in terms of adaptation to climate scenarios and temperature. The mimicking of natural elements is termed biomimicry. A recent example of biomimicry is the solar shade design used in the courtyard of Masdar's city center.

Masdar city's courtyard is being designed to take full advantage of the natural climate and conditions of its desert setting. The center of Masdar has been designed by LAVA architects with sunflower-like structures, referred to as solar sun shades, that open and close like a sunflower, providing shade during the day and then closing at night to

Figure 6.24 The Solar Sun Shades at Masdar City Center's Courtyard during the Night.

Source: LAVA: Laboratory for Visionary Architecture, n.d.

let the heat escape and to let the area naturally cool with the night air. The architects have called these shades "petals from heaven", which along with other features of the city, will draw residents to this oasis of life. The solar sun shades are illustrated in Figure 6.24. Masdar city uses traditional narrow streets for shade with a diagonal layout designed to capture the evening winds providing cool corridors through the city (Figure 6.25). The temperature of the city is predicted to be much lower than the surrounding areas.

6.3.7. *ACROS Fukuoka Prefectural International Hall*

The Asian Crossroads Over the Sea (ACROS) Prefectural International Hall is a mixed commercial-public building located in Fukuoka, Japan. Along with the Prefectural International Hall, it hosts an exhibition hall, a museum, a 2000-seat proscenium theatre, conference facilities, 600,000 square feet of government and private offices, a large underground parking area and retail spaces. It was built on the last piece of green space in Fukuoka, the Tenjin Central Park; therefore, the main idea was

Figure 6.25 Narrow Streets in Masdar.

Source: Peter Newman.

to create new public land equal to that lost. This was done through the creation of a green roof system facing a park into the design. This project contributed to both the advancement of culture for residents of Fukuoka and the vitalization of the local community.

ACROS resembles a natural mountain from both its appearance and the system it provides. Its 'Step Garden Roof' is home to 35,000 plants and 76 species, supporting many birds and insects. The roof captures rainwater runoff and directs it to waterways partially treating it along the way. The rainwater is used to water the plants and to mask the ambient noise of the city. The park is used for many festivals, particularly, the famous Don Taku Festival. The roof garden utilizes many local plants, chosen for their sturdiness to local typhoon conditions. The park offers an escape from the congested city that surrounds it with people using the area for meditation and relaxation.

Due to the garden roof, the temperature of the building is 15 degrees less than the ground temperature, thus generating less power consumption. The building also utilizes natural daylight, saving the energy and creating a natural ambience inside this 15-story building.

6.4. Conclusion

The cases presented here from across Singapore and various Asian cities illustrate the biophilic city concept. They show examples of working towards a greener, more hopeful future. It is interesting to note that these examples usually stem from small steps which through time become significant. One small step in greening the grey city progressively influences the surrounding areas, creates more biodiversity, cools the environment, removes legacy carbon, increases food security and enhances the health and productivity of the people. These developments help propel a greener future by bringing nature back into the city, meanwhile also enhancing the economy and social aspects of the community. They are therefore both symbolic and real first steps.

While Singapore may be at the forefront in Asia of creating a biophilic city, particularly due to its commitment to creating a 'City in a Garden', it still lacks the urban farming aspect necessary to increase its food security. Green landscapes can be more than just visually-pleasing landscapes. A productive green landscape enhances the economy and enriches the community spirit. Food security will be an important part of every city's future and can be built into the biophilic rejuvenation of public spaces and buildings.

Meanwhile, as cities copy each other, most Asian cities have understood and have seized upon the opportunity to be a biophilic city. Cities such as Tokyo, Seoul and Shanghai have all started to connect their parks (copying Singapore) to create an integrated park network within the city. The next step will be to see how dense buildings will be transformed as Singapore is doing.

CHAPTER SEVEN

The Eco-efficient City

7.1. Introduction

The eco-efficiency concept sees cities as complex metabolic systems with material and energy flows and cycles. Importantly, the concept enables places to re-envision elements, such as wastes, that are generally viewed as negative outputs as productive inputs and resources (Newman *et al.*, 2009).

In Asia, the eco-efficiency concept is not a new idea. Many projects have tried to implement the concept at varying scales. Some projects have been planned with this concept in mind since the beginning and others have adopted the concept during their life cycle to improve the existing environment along with managing waste and wastewater issues. The case studies provided in this chapter look at Singapore at a national scale, the new development of the New Songdo City in South Korea and Beijing's Changxindian Low-carbon Community as regional scale examples, Nanjing Eco High-tech Island, Kawasaki Eco-town, Surat Thani Biomass Power Generator, Bekasi City in Indonesia, El Nido Resort in the Philippines, Manav Sadhna Activity Center in Ahmedabad, India, and Trashion in Indonesia as examples of eco-efficient examples on a local scale.

7.2. Singapore

As a city-state with an area of 710 km^2 and a population of 5,076,000 people, limiting the waste output and minimizing the use of resources are very critical aims for Singapore's sustainable growth and development. This is not an unusual challenge in Asia and is one of the reasons that eco-efficiency has become so important in Asian cities. At the Asia 3R Conference in 2006, the Singapore National Environment Agency and the Ministry of the Environment and Water Resources highlighted that there were four main challenges in delivering Singapore's waste management (Asia 3R Conference, 2006). The first and the most critical challenge was the small land area of Singapore limiting the available area for landfill. The second challenge was economic growth, which has very strong connections with the amount of industrial waste. The third challenge was the population growth and its effect on the amount of domestic waste produced. The final challenge was the rising affluence of the Singapore citizens which leads to higher expectations in lifestyle and consumption patterns (Ministry of the Environment and Water Resources, 2007).

The change of lifestyle and consumption patterns over the last decade in Singapore has increased the amount of domestic waste produced. In a market survey conducted in 2001, the total domestic waste disposed per capita was 1.14 kg per person per day. Compare this rate to the total domestic waste disposal per capita of 0.75 kg per person per day in 1980 (Waste Management and Recycling Association of Singapore, 2001). This is quite a dramatic increase. In 2011, the total waste output of Singapore was 6,898,300 tonnes with a recycling rate of 59% (Figure 7.1; National Environment Agency, 2012).

The concept of the eco-efficient city aims to reduce the ecological footprint of the city by reducing waste outputs as well as the resource inputs to a city by 'closing the loop' on this metabolic flow (Newman *et al.*, 2009). This concept has been successfully implemented in Singapore's waste management. Various strategies and programs from statutory boards and from the private sector have helped Singapore to reduce the amount of waste disposed and increase the rate of recycling. Policies and regulations, particularly the Environment Public Health Act, the Code of Practice on Pollution Control, and the Code of Practice for Licensed General Waste Collectors, amongst others, have been used to provide a legal basis for the implementation of Singapore's waste management (National Environment Agency, 2012).

7.2.1. *Singapore's Solid Waste Management Strategies*

Singapore's solid waste management strategy is focused around three key concerns (Ministry of the Environment and Water

Waste Statistics and Recycling Rates for 2011				
Waste Type	Waste Disposed (in tonnes)	Total Waste Recycled (in tonnes)	Total Waste Output (in tonnes)	Recycling Rate (%)
Food Waste	605,800	69,700	675,500	10%
Paper/Cardboard	603,200	765,000	1,368,200	56%
Plastics	656,000	77,000	733,000	11%
Construction Debris	12,600	1,191,100	1,203,700	99%
Wood/Timber*	97,400	176,500	273,900	64%
Horticultural* Waste	149,800	89,000	238,800	37%
Ferrous Metal	67,600	1,171,600	1,239,200	95%
Non-ferrous Metals	14,500	102,800	117,300	88%
Used Slag	5,700	335,900	341,600	98%
Sludge	152,900	0	152,900	0%
Glass	51,400	21,400	72,800	29%
Textile/Leather	113,700	17,300	131,000	13%
Scrap Tyres	3,700	18,300	22,000	83%
Others (stones, ceramics and rubber)	325,200	3,200	328,400	1%
Total	2,859,500	4,038,800	6,898,300	59%

Figure 7.1 Singapore's Waste Statistics and Recycling Rates for 2011.
Note: *Includes 66,600 tonnes used as fuel in biomass power plants.
Source: National Environment Agency, 2012.

Resources, 2007):

1. Waste minimization at the source;
2. Recycling to reduce waste disposed at incineration plants and landfills; and
3. Volume reduction of waste disposed through incineration.

As part of this, the National Environment Agency has worked with industry players on a Voluntary Packaging Agreement to reduce the amount of packaging waste generated in order to encourage waste minimization at source. The Agreement, which commenced on 1 July 2007, provides a platform and structure for industries to collaborate with the government to reduce packaging waste over a five-year period. Of this, about one third was food packaging waste. From 1 July 2007 to

30 September 2009, the Agreement specifically targeted food and beverage packaging. This type of packaging constitutes more than 50% of all household packaging waste. The Agreement has resulted in reducing Singapore's total domestic waste disposed per capita from 0.88 kg per person per day in 2007 to 0.84 kg per person per day in 2009, with domestic waste forming 58% of all waste disposed in Singapore in 2009. Since 2009, the focus of this Agreement has been extended to cover all types of product packaging, including detergents, toiletries, personal care products and household products (National Environment Agency, 2012).

Recycling in order to reduce the amount of waste being disposed by incineration or landfills plays a key role in Singapore's efforts to close the waste loop. Waste-to-energy (WTE) plants offer a solution to reduce waste volume efficiently and to conserve landfill space. Singapore has four WTE plants: Tuas, Senoko, Tuas South and an offshore sanitary landfill, Semakau Landfill. The 3Rs (reduce, reuse, recycle) also play a crucial complementary role to WTE plants by preventing waste at the source and by cutting down on the amount of waste sent to the disposal sites and enabling resources to be recovered (National Environment Agency, 2012). Singapore's National Recycling Program was launched in April 2001. Under this program, Public Waste Collectors (PWCs) licensed by the National Environment Agency are required to provide recycling collection to all households in both public and private housing. The Schools Recycling Corner Program was launched in September 2002 with the aim of educating students about the need to recycle. A recycling program for JTC Industrial Estates was launched in November 2003 which collected waste for reuse along with the collection of recyclables (National Environment Agency, 2012).

Recyclables collected under the National Recycling Program are sent to Material Recovery Facilities (MRF) where they are sorted into more specific waste streams (e.g. paper, plastic, glass, clothing, ferrous and non-ferrous scraps) before being sent to a recycling company where they are recycled into a resource. Food waste is converted into compost and biogas using a bio-methanisation process. Biogas is used as a fuel source for supplying electricity to the power grid. Plastic waste is recycled into plastic pellets, which are further supplied to manufacturers for use as feedstock in the production of other plastic products. Wood waste is converted into 'technical wood', which is used as raw materials for products such as tables, doors, flooring and building materials. Construction and demolition waste are turned into eco-concrete and other materials (National Environment Agency, 2012). In 2011, used slag, construction debris and ferrous metal had the highest recycling rates, with sludge, plastics and other waste types (stones, ceramics and rubber) having very low recycling rates. Singapore's total recycling rate in 2011 reached 59%, with a total recycled waste weight of 3,485,200 tons (Figure 7.1).

Most non-recycled waste goes to incineration plants with some going to landfills. In land-scarce countries like Singapore, incineration is a cost-effective disposal solution, reducing waste volumes going to landfills by as much as 90%. Incineration, however, produces flue gases which can cause air pollution that need to be cleaned before they are released into the atmosphere. Currently, Singapore has four WTE incineration plants with a total capacity of up to 7,600 tonnes of waste per day. The Tuas South Incineration Plant has the largest capacity at 3,000 tonnes of waste per day and a power generation capacity of 80 MW. The plant cleans flue gas before releasing it using a three-part system. The flue gas is first passed through an electrostatic precipitator which removes 95% of the fly ash and dust particles. This is then passed through a catalytic fabric filter system to remove the remaining particles. Furthermore, a hydrated lime powder is mixed with the flue gas to reduce the acidity of the gas. This is then dispersed into the air through two 150 m high chimneys. The heat produced by the incineration process is used to generate electricity, 20% of which powers the plant, with any excess sold to Singapore Power. The plant uses recycled industrial wastewater in its boilers along with captured exhaust steam. Scrap steel extracted from the incineration ash is sold to a local steel mill. The four WTE plants produce 2–3% of the national electricity demand.

Non-incinerated waste and incineration ash are disposed of at the Semakau Landfill. The Semakau Landfill was created on reclaimed sea space and has been in operation since 1999. The reclamation cost S$610 million (approximately US$399 million). The landfill highlights Singapore's commitment to balancing urban development and nature conservation — 13 hectares of mangroves were replanted to replace some of those that had to be removed during the construction and the area is now home to a wide variety of plants and animals, including seven species of sea grass, hard corals and marine life (Tan, 2012a; 2012b). The landfill has a capacity of 63 million m^3, enabling it to meet Singapore's anticipated landfill needs beyond 2040 (Waste Management World, 2011). Parts of the island are open to the public for recreation. The combination of recycling and incineration has reduced the amount of waste that goes into landfills by as much as 95%.

7.2.2. *Singapore's Wastewater Management*

Singapore has had to overcome severe water shortages, flooding and high pollution in its rivers prevalent in the 1960s and 1970s. To overcome this, and to enable Singapore to be sustainable in water, the Singapore Government has been investing in research and technology around sustainable water management. This investment has enabled the nation to develop a diversified and sustainable water supply. This supply is from four different sources: water collected from local catchment areas, imported water, reclaimed water (NEWater) and desalinated water. Collectively these sources are known as the 'Four National Taps'. Singapore has managed to close the water loop by creating an integrated system that maximizes the efficiency of all of the water sources (Public Utilities Board, 2010).

Singapore started reclaiming used water for non-potable use in 1972 in order to conserve potable water and close its water loop. In 2002, the Public Utilities Board (PUB) embarked on the NEWater initiative. NEWater reclaims wastewater and purifies it using advanced membrane technologies, microfiltration (MF), reserve osmosis (RO) and UV disinfection. This water can then be used in industries and for air conditioning in commercial and institutional complexes (PUB, 2012). Furthermore, a small percentage of NEWater is used in reservoirs where it is further processed and used for drinking water. Singapore currently has four NEWater factories: Bedok (2003), Kranji (2003), Ulu Pandan (2007) and Changi (2010). NEWater currently supplies 30% of Singapore's water needs, with plans to triple this capacity by 2060. The NEWater reclamation won the Stockholm Industry Water Award in 2007.

Singapore uses two systems to collect rainwater and stormwater. Rainwater is collected through a network of drains, canals, rivers and stormwater collection ponds before it is channeled into one of Singapore's 17 reservoirs for storage. Singapore is one of the few countries that collect stormwater on a large scale for use in its water supply. Combined with Marina Reservoir, which is the most urbanized catchment, the two newest reservoirs, Punggol and Serangoon Reservoirs, increased the water catchment areas from half to two-thirds of Singapore's land area (PUB, 2010).

Singapore has become an iconic example of the eco-efficient city concept in its water supply through the reuse of sewage and stormwater as well as its recycling of solid waste.

7.3. New Songdo City, South Korea

New Songdo City is a master-planned city currently under construction on the Incheon waterfront, 65 km west of Seoul in the Republic of Korea (South Korea) (Kon Pedersen Fox Associates, 2003). The city is connected to Incheon International Airport by a 12.3 km highway bridge. The city is being built on land reclaimed from the ocean and will be 6 km^2 in size and will form part of the Incheon Free Economic Zone (IFEZ). The IFEZ aims to create hubs of various economic zones (e.g. international business, leisure and tourism). The IFEZ, when completed, will feature residential and business districts, an international business center, shopping and entertainment centers along with the necessary services. The project is expected to take ten years to develop and is estimated to cost an excess of $40 billion. It is due for completion in 2015 and will have 80,000 apartments, 4,600,000 m^2 of office space and 930,000 m^2 of retail space.

Regional economic pressure has been one of the major reasons behind the development of New Songdo City. In the past 15 years, Korea has lost more than 50% of its manufacturing business to China and other Asian countries where the labor costs are less expensive and the skills are improving. In order to recapture the country competitiveness, Korea has developed the IFEZ. This IFEZ shifts the orientation of national economy to financial and logistical service industries.

These zones allow for favorable business practices that are meant to encourage international trade and foreign investment.

A typical Free Economic Zone (FEZ) is driven primarily by economics failing to address social and environmental concerns. At a microeconomic scale, New Songdo City aims to be an international trading hub, broadening and increasing the vitality of Korean business. In order to achieve such an ambitious goal, New Songdo City raises the standards of typical FEZ development by aiming to attract global business through the development of a sustainably-designed built environment enriched by cultural, social, educational and recreational amenities. In this way, Songdo is seen by South Korea to be part of its green economy plan. In August 2008, the president of South Korea, Lee Myung-Bak, announced that green growth would be the basis for all future South Korean developments. Since then a number of green economy initiatives have been introduced including a National Strategy for Green Growth (2008), a National Greenhouse Gas Emissions Reduction Target (2009) to reduce emissions by 30% by 2020, the Green New Deal (2009), a Five-Year Plan for Green Growth (2009) which envisions South Korea as one of the top seven green economies in the world by 2020 and a Framework Act for Low-carbon Green Growth (2010) which enables the government to intervene in the private market if businesses fail to promote green, low-carbon growth (United Nations Environment Programme, 2010; Young, 2012). South Korea approved a 'cap-and-trade' carbon reduction system in May 2012 (Han, 2012). New Songdo City is an integral part of the green economy plan.

Rapid urbanization and the use of non-renewable resources have been considered the main concerns that need to be addressed in this new green economy industrial area. From 1960 to 2000, Korea went from an 80% rural population to an 80% urban population. The rapid urban population growth is increasing water and air pollution while diminishing the quality of the land, aggravating already chronic traffic congestion and placing increasing burdens on energy resources. Energy consumption in South Korea has increased substantially with energy use increasing by 8.2% per year and energy-related CO_2 emissions increasing by 7.0% per year between 1990 and 2005 (Oh *et al.*, 2010). South Korea is ranked within the top ten highest CO_2 polluters in the world. These concerns formed the basis of a 2006 Environmental Opportunity Charrette. This charrette identified ways to improve environmental performance and quality of life in New Songdo City (Council on Tall Buildings and Urban Habitat, 2008). The charrette focused on developments that considered long-term sustainability in every decision and that would be adaptable to changing needs. Specifically, the focus was on site planning, energy use, building designs, water consumption, transportation, the material choices and waste production and the impacts of development. The charrette also focused on 'soft' design elements to enhance social sustainability, encourage alternative forms of transport and a vibrant public realm, looking at the principles of transit-oriented developments (TODs), new urbanism and those of traditional Korean *'dongs'* or neighborhoods (Council on Tall Buildings and Urban Habitat, 2008).

New Songdo City aims to become an international destination known for providing a high quality of life and having a vibrant public realm rich in cultural and recreational attractions. Particularly, one of the goals is to enable South Korea to adapt to the changing global economic environment, to enable job creation and provide modern, energy-efficient dwellings. One of the aims is that all of the 350 buildings to be developed would receive the Leadership in Energy and Environmental Design (LEED) Green Building certification. The future viability and health of the city, along with minimizing South Korea's impact to global climate change, is at the forefront of design considerations. New Songdo City is designed around six core goals:

1. Open and green spaces;
2. Transportation;
3. Water consumption, storage and reuse;
4. Carbon emissions and energy use,
5. Material flows and recycling; and
6. Sustainable city operations.

Some of these features are described in Box 7.1. The city has been designed using an orthogonal pattern of streets interlaced with open spaces, walkways and canals (Figure 7.2). New Songdo City will have 515 acres of landscaping and open spaces, comprising 34% of the total land area of the development. It is hoped that these facilities, combined with the mixed land uses in the area, will encourage walking and cycling.

Box 7.1 Songdo City's Six Core Design Goals and Some of the Measures to Reach Them.

New Songdo City's Six Core Design Goals.

1. *Open and green spaces: Providing connected and amenity-rich open and green spaces.*
Some measures:

- 40% open space (600 acres).
- Central Park (100 acres).
- Open and green spaces connected by walking and cycling corridors.
- Design maximizes optimal access to sunlight and views.
- Ecological diversity utilizing native and water-efficient species.

2. *Transportation: A pedestrian-oriented urban center, encouraging non-polluting, quiet and city-friendly travel, and discouraging private car use.*
Some measures:

- The Incheon subway line will run through the center of Songdo IBD.
- Expanded bus service.
- 25 km network of bicycle lanes.
- 5% of parking capacity within each project block will be reserved for fuel-efficient and low-emitting vehicles. Office and commercial blocks will reserve an additional 5% of parking for carpool vehicles.
- Charging stations for electrical vehicles will be integrated into parking garage designs.

3. *Water consumption, storage and reuse: Sustainable water storage and use will be maximized.*
Some measures:

- Rainwater and stormwater harvesting and use.
- Treatment of grey and black water for appropriate reuse.
- The Central Park canal will use seawater.
- Use of efficient landscape design and irrigation systems.
- Use of efficient plumbing fixtures.
- Vegetated green roofs.

(*Continued*)

Box 7.1 *(Continued)*

4. Carbon emissions and energy use: Provision of energy-efficient and comfortable buildings and the efficient use of energy city-wide.

Some measures:

- All New Songdo International City Development projects will be built to or above the standards of the American Society of Heating, Refrigerating and Air Conditioning Engineers (ASHRAE).
- A central, city-wide co-generation facility fueled by natural gas providing power and hot water.
- Energy-efficient LED traffic lights and energy-efficient pumps and motors.
- A centralized pneumatic waste collection system to collect wet and dry waste.

5. Material flows and recycling.

Some measures:

- Recycling of construction waste (target of 75%).
- The use of local and/or recycled materials will be maximized.
- The use of low- (or lower) carbon and low-VOC materials will be maximized.

6. Sustainable city operations.

Some measures:

- Sustainable procurement goals and recycling guidelines integrated into the operational structure of the city.
- Management and maintenance contracts will mandate environmentally-friendly products (low or zero VOC, eco-label, 'Good Recycled' designations or equivalent).

Source: Songdo IBD, 2012.

New Songdo City is another exemplary case of how government vision has been implemented for the development of a more eco-efficient city. There is a strong marketing push behind Songdo; however this example demonstrates how eco-cities have become a new trend of the emerging green Asian city. None of the technologies applied here are new, but the effort of planning and developing an integrated city from the inception has allowed eco-efficiency to be built in from the start. In this way

Figure 7.2 Songdo City Perspectives.

Source: Songdo IBD, 2012.

South Korea is able to demonstrate its leadership in this new green economy.

7.4. Beijing Changxindian Low-carbon Community Concept Plan

A low-carbon community concept plan was developed by Arup in collaboration with the Beijing Municipal Institute of City Planning and Design (BMICPD) for a 500 hectare site within the Fengtai District of Beijing, approximately 17 km from central Beijing (Arup, 2009). The Beijing Changxindian Low-carbon Community Concept Plan was designed to incorporate sustainable community design principles, eco-efficiency and had 20 quantitative performance indicators for minimizing and measuring the use of water, energy and the production of carbon emissions through statutory zoning. The plan was for a mixed-use concept with a residential area to serve approximately 70,000 people along with public open spaces and commercial and research industries all served by a light rail. Part of the plan was to develop 'low-carbon zoning codes' that are implementable as statutory zoning plans. The plan was compared to a 'business as usual' development using an ecological footprint measure and an integrated resource management framework, determining that residents of the proposal would have an ecological footprint 27% less than those in the 'business as usual' scenario. All of the performance targets and technical requirements were modeled on a site, block and whole area basis to ensure that the sustainability targets were being met and also in order to develop site-specific low-carbon zoning codes that could be embedded into the existing regulatory codes. This pilot project was a first for China and aimed at overcoming some of the regulatory barriers to low-carbon, eco-efficient developments.

7.5. Nanjing Eco High-tech Island

The Nanjing Eco High-tech Island (Nanjing Eco-city) is a Chinese eco-city being developed in joint partnership with the government of Singapore. The goal of this eco-city is "to establish a platform for the sustainable development of high-tech, smart industries under an ecologically conscious environment" (Alusi *et al.*, 2011, p. 11). In November 2007, a bilateral platform, the Singapore–Jiangsu Cooperation Council, was launched to promote collaboration between Singapore and Jiangsu on the eco-city project. International Enterprise (IE) Singapore signed a Memorandum of Understanding (MOU) with the Nanjing Municipal People's Government, agreeing to jointly undertake a feasibility study of the Eco High-tech Island. The final agreement was formed in 2009, and the city is targeted for completion in three phases by 2020.

The Nanjing Municipal Government procured the land, with the real estate being developed by Singapore Intelligent Eco Island Development and Nanjing Jiangdao Investment and Development Company. Singapore Intelligent Eco Island Development is a private joint venture company formed by Yanlord Land Group (the real estate developer), Sembcorp Industrial Parks (the industrial development and engineering wing) and Surbana Land (who will look at the design and architecture of the city) (Yanlord Land Group, Sembcorp Industries and Sembcorp Parks Management, 2009). The Nanjing Eco-city has 1,280 acres of land to be developed.

The Nanjing Eco-city will be based on high-tech and smart industries in an ecologically aware environment and follows the movement to have eco-industrial parks and a circular economy policy in China (see Box 7.2). Some of the eco-efficiency features that Nanjing Eco-city will have are: optimizing the use of water and power; creating an emission-controlled and smart wastewater management system; a waste-to-resource approach; an eco-trip transportation network; and sustainable industry development (Alusi *et al.*, 2011). Eco-efficiency is obviously a major policy for China and Nanjing Eco-city will demonstrate its application in a real city.

Box 7.2 Eco-industrial Parks and the Circular Economy Policy in China

The circular economy policy adopted in China in 2002 and the Circular Economy Promotion Law enacted in 2009 are concerned with meeting environmental, economic and social needs, improving resource productivity, optimizing material use and eco-efficiency and are based on ideas of a closed-loop material flow. In China this is taken to the level of the whole economic system (Geng and Doberstein, 2008; Geng *et al.*, 2010). The idea focuses on industrial ecology or eco-industrial development (EID) where efficiencies between industries are utilized to minimize the use of energy, water, materials and the sharing of common infrastructure. This focus arose due to the environmental degradation that has resulted from China pursuing rapid industrial production since the 1980s (Geng and Zhao, 2009). A main component of EID is that the waste by-products produced by one industry or activity can be used as a material or resource in another industry or activity.

The circular economy model is implemented in three circles:

- The *first circle* is focused on micro-level initiatives particularly on cleaner production following the enactment of the Cleaner Production Promotion Law in January, 2003.
- The *second circle* is the meso-level and involves cooperation and efficiencies between businesses through the creation of eco-industrial parks (EIPs) where businesses are able to utilize industrial by-products such as heat energy, wastewater and manufacturing wastes. In China industrial parks usually have a residential component along with the industrial so this concept also includes safe residential measures.
- The *third circle* is the macro or social level and involves the development of 'eco-cities' and eco-districts. This level focuses on efficiencies in production and also at reducing consumption levels within the society.

Sources: Geng and Doberstein, 2008; Geng and Zhao, 2009; Geng *et al.*, 2010; Xi *et al.*, 2011.

7.6. Kawasaki Eco-town, Japan

Eco-towns in Japan are recycling-oriented industrial clusters, initiated in 1997 through a subsidy system established by the Ministry of Economy, Trade and Industry (METI) and the Ministry of Environment (MoE) in order to facilitate the development of the recycling industry (Global Environment Centre Foundation, 2005; Norton, 2007). The aims of the project were to lessen the waste going to landfills (which Japan had a serious shortage of), to revive the local economy, to assist declining industries and to implement the zero-emission concept. In 2006, there were 26 eco-towns in Japan (Figure 7.3), with Kawasaki Eco-town being one of the first developed in Japan.

The zero-emission concept, developed by the United Nations University in 1994, focuses on reducing the amount of waste generated by industries and companies and to make sure that waste is properly recycled in collaboration with other industries. The zero-emission concept follows four principles:

1. Gross input must equal gross output with the aim of getting the waste output close to zero;
2. Reduction in greenhouse gases and environmental burdens;
3. Promotion of energy-saving measures; and
4. Collaboration among collective industries and administrative districts (Global Environment Centre Foundation, 2005).

Eco-town projects in Japan	
Aichi Prefecture	Kochi City (Kochi Prefecture)
Akita Prefecture	Minamata City (Kumamoto Prefecture),
Aomori Prefecture	Naoshima-cho (Kagawa Prefecture)
Chiba City (Chiba Prefecture)	Okayama Prefecture,
Ehime Prefecture	Omuta City (Fukuoka Prefecture)
Gifu Prefecture	Sapporo City
Hiroshima Prefecture	Suzuka city (Mie Prefecture)
Hokkaido	Tokyo
Hyogo Prefecture	Toyama City (Toyama Prefecture)
Iida City (Nagano Prefecture)	Uguisuzawa-cho (Miyagi Prefecture)
Kamaishi City (Iwate Prefecture)	Yamaguchi Prefecture
Kawasaki City	Yokkaichi City (Mie Prefecture)
Kitakyushu City,	

Figure 7.3 Eco-town Projects in Japan, 2006.

Source: Fujita, 2006.

Eco-towns implement the zero-emission concept through urban planning and environmental management (Global Environment Centre Foundation, 2005).

Kawasaki Eco-town is an industrial area in Kawasaki City, located in the coastal expanse of the Tokyo Metropolitan area. Kawasaki Eco-town is the gateway to Japan with its accessibility by land, sea and air transportation. The 2,800 hectare industrial site was initiated in 1997 with the aim to create a resource-recycling society and develop a zero-emission industrial park. The total budget of the project was about 25 billion *yen*, provided as a subsidy by the national and city governments. The site is on a land reclamation area, the Kawasaki Coastal Area, the reclamation of which began in the Meiji period (1869–1919). The full-scale industrialization of the area began in 1913. After World War II, Japan enjoyed high economic growth, and large-scale industrial plants began concentrating in the Kawasaki Coastal Area. The accelerated industrialization of the coastal area resulted in a dramatic increase in fuel and water consumption and also caused air and water quality degradation due to the discharge of pollutants from factories. This problem accumulated and in the 1990s, Japan faced two serious problems: environmental issues and the breaking of the economic bubble. Citizens began to protest particularly about the pollution.

The zero-emission concept was developed to respond to this problem and also as a way of developing new industries based on zero-emission technology. Much research was done on how to apply the concept to various industrial areas. This led to the redevelopment of the industrial area in Kawasaki City to an eco-town. The goals of the Kawasaki Eco-town are as follows:

- Individual industrial firms strive to reduce all types of environmental impacts throughout the entire process of their activities from manufacturing to waste disposal.
- Companies are to cooperate jointly to recycle products and use recycled products.
- Environmental research is to be continued in order to sustain growth in the Environmental Recycling Zone. Environmental research focuses on the use of factory waste heat and the use of product recycling systems as materials for various businesses.
- The knowledge of the eco-town will be disseminated to other parts of Japan and to foreign countries.

The eco-efficiency effort in Kawasaki Eco-town is intended to create a circular, closed-loop system. The recycling facilities were constructed to recycle waste from the nearby manufacturing firms specifically from the steel, non-ferrous metal, cement, chemistry and paper manufacturing firms. The facilities also serve as a recycling center for the two major nearby cities of Tokyo and Yokohama. The mutual use of the by-products of processing and the use of waste as raw materials has been promoted in the newly-built recycling plants and companies in Kawasaki Eco-town, for example:

- The Showa-Denko (chemical) company provides ammonia produced in the manufacturing process to other enterprises.
- Corelex Co. Ltd. (paper) generates incinerated ash during the process of transforming used paper into toilet paper and provides this by-product as a raw material to cement companies. Furthermore, Corelex Co. Ltd. utilizes the surplus electricity of JFE Steel Co. Ltd. and uses recycled water treated in a municipal wastewater treatment plant.
- Nihon Yakin Co. Ltd. uses raw materials generated by JFE Steel Co. Ltd. from the recycling process of electric appliances to produce a special alloy.

Kawasaki Eco-town, although initiated by national and city governments, gained much support and collaboration from various stakeholders by enabling them to take on different roles within the project. Recent collaborations amongst industries within Kawasaki Eco-town are the development of a large-scale energy saving project where Tokyo Electric Power Company (TEPCO) supplies steam from its thermal electric power plant to surrounding industries and a 20,000 kW joint mega-solar power generation project between Kawasaki City and Tokyo Electric Power Co. that commenced operation in 2011.

The Kawasaki Eco-town has demonstrated remarkable improvement of environmental quality, particularly in improving air quality. Research by the Center for Regional Industrial Symbiosis Research at Toyo University estimates that the recycling of waste reduced CO_2 by 160,000 tons per year. The Kawasaki Eco-town case study illustrates a good example of what can happen with government commitment, collaboration among industries and with community support to reduce carbon production and to reduce the negative impact on the environment caused by industrial areas. The eco-efficiency concept has driven the transformation of the industrial area from a conventional industrial framework which maintains that the industry should be separated from other land uses and other industries, to a framework based on zero-emissions and a closed-loop waste system. This closed-loop system requires carefully clustered industries so that there can be mutual use of by-products and wastes, with wastes from one industry used as material sources in other industries within the designated area, whilst still enabling the industrial park to be economically and socially viable. The Kaswasaki Eco-town is an important demonstration of how an old industrial area can be retrofitted to enable eco-efficiency in the same way that new areas can be planned as a closed-loop system.

7.7. Surat Thani Biomass Power Generator, Thailand

The Surat Thani biomass power generator is a power generation project that uses empty fruit bunches (EFB), a waste product of the palm oil milling process, as a primary fuel for power generation (United Nations Framework Convention on Climate Change (UNFCCC), 2006). The project is part of the Thai government's effort to promote the domestic production of biofuels in order to cut down the use of imported oil by 1,000,000 kiloliters (kl) in the next five years (Asia Biomass Office, 2008). This would save up 42 billion bahts. Furthermore, the production of biofuels would increase the income of farmers compared to the farming of conventional agricultural products.

The domestic production of primary energy in Thailand during the 2008 financial year came up to 62,695 kilo-ton oil equivalent (ktoe), of which renewable energy was 19,295 ktoe (30.8%) and biofuel was 626 ktoe (1%) (Asia Biomass Office, 2008). Thailand is the biggest biofuel producer amongst the Southern Asian countries, producing 300,000 kl of bioethanol and 400,000 kl of biodiesel in the 2008 financial year (Asia Biomass Office, 2008).

The Surat Thani Biomass Power Generator is located in Amphur Phun-Phin, Surat Thani Province, about 700 km from Bangkok. The project is a collaboration between Surat Thani Green Energy Co. Ltd. (SGEC), a private Thai company, and with Agritech Marketing Co. Ltd. (AMC), a recycling company authorized by the Japanese Ministry of Environment, and Ministry of Agriculture, Forestry and Fisheries. The project was initiated in 2005 and commenced commercial operations in 2007 (UNFCCC, 2007).

In addition to providing renewable energy, the Surat Thani Biomass Power Generator will improve the disposal of agricultural waste, adding to its application of the eco-efficiency concept. Before this, EFBs has no commercial value and was left to decay at solid waste disposal sites. EFBs used by the Surat Thani Biomass Power Generator are from six mills within the province. The six mills will be able to supply 350.6 kilo-ton per year of EFB, which is more than enough to fulfil the 252.6 kilo-ton per year requirement to generate 70,168 MWh of electricity per year (UNFCCC, 2007). The electricity produced by the Surat Thani Biomass Power Generator will be sold back to the local electricity grid under the Small Power Producer Program.

The social benefits for the local community will be the creation of new job opportunities, such as the palm oil manufacturer, truckers, power plant operators, and construction workers.

In agro-industry countries such as Indonesia, Vietnam and Thailand, the residues of these industries such as rice husk, bagasse, oil palm residue and the empty fruit bunches often become a major problem, especially in air and river pollution. These are considered to be waste products and are often burned, causing pollution and CO_2 emissions as well as disposal into rivers. Furthermore, poorer households often burn this biomass waste in inefficient stoves, resulting in indoor smoke pollution. This indoor smoke pollution from burning biomass is estimated to cause the death of approximately 2 million people each year worldwide. Using this biomass in larger, more efficient power plants and converting it into electricity instead of burning it directly in stoves is a much more beneficial outcome for both the environment and the communities. The Surat Thani Biomass Power Generator plant demonstrates that eco-efficiency can work in lower-income areas of Asian settlements and their bioregions, as do the projects outlined below.

7.8. Bekasi City, Indonesia

Bekasi is a city in West Java, Indonesia, located on the eastern border of Jakarta in the Jabodetabek Metropolitan region. It has an area of 210.49 km^2 and has a population of 1.9 million (in 2005) and a population density of 9,471 people per km^2. Bekasi serves as a commuter city for Jakarta, although it has notable trade, business and processing industries. In 1989, one of the districts of Bekasi, Bantar Gebang, was planned to be the final collection point for all the waste from East Jakarta. However, in reality, the landfill was used to collect waste from all parts of Jakarta. Consequently, each day the incoming waste exceeded the capacity of the landfill.

The vast growth and activities of Jakarta has resulted in increasing amounts of waste and associated problems. Based on the data from the local government, in 2007 the total amount of waste dumped per day was about 27,654 m^2 (equivalent to 6,914 tons). At the moment, Jakarta only has one final waste dump/landfill at Bantar Gebang, Bekasi City, with an area of 108 hectares with no waste processing or gas control being applied. The facilities utilized open dumping: dumping waste without any processing. This disposal technique did not apply any proper waste processing technology and has harmful disadvantages causing permanent pollution to the land and water as the waste disposed of in this manner generates leachate liquid that is very dangerous as well as methane which is vented to the atmosphere with high climate impacts.

Moreover, the open dump is located near residential environments. The gas that resulted from the decomposition of organic materials is harmful to the neighborhood as it contains hydrogen sulfide.

Green Indonesia, a non-profit independent organization, held a rally on November 10, 2008, with thousands of activists, student organizations, intellectuals, volunteers, social workers, academics and young professionals gathered with the aim of increasing the quality of human living conditions. Bekasi City was chosen because of the social stigma attached to this city as 'Beast Cities' and 'Trash City' because of the problem of the garbage landfill. The city was literally the waste container for the city of Jakarta (Green Indonesia, 2011).

Following this rally, in 2009 Green Indonesia launched several activities around community-based development using a 'Bottom-up Planning' approach. The activities focused on the following programs:

1. Planting 100,000 trees in Bekasi City through an 'adopt a tree' method.
2. A public campaign to implement the 5Rs of waste management (Reduce, Reuse, Recycle, Rot and Respect).

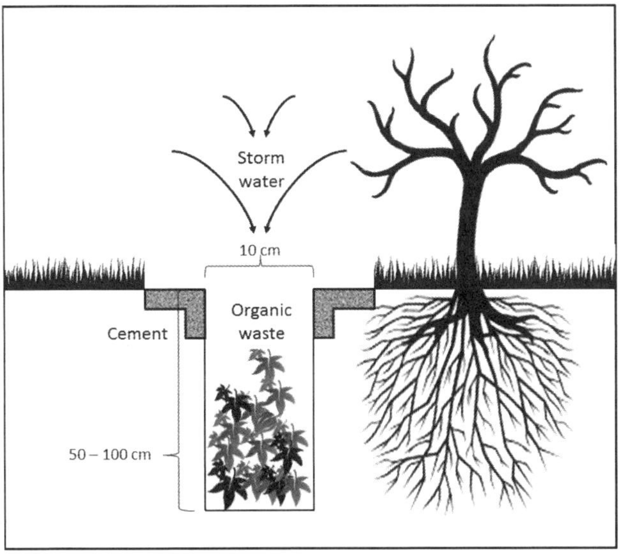

Figure 7.4 Biopores.
Source: Prepared by the authors.

3. A water conservation and flooding program and public campaign that aims to make 1 million *Biopores*, an infiltration well that increase the absorption capacity of the soil (Figure 7.4), and the reusing of organic waste as a compost.
4. Programs providing 1 million *Trembesi (Albizia Samanea)* tree seedlings per year for the Reforestation Program in Indonesia.
5. Facilitating a 'Green School' implementation program at international standards around Indonesia.

Bekasi City was selected as a pilot development program with the vision *'My Bekasi is Green, My Bekasi Flood Free and My City Bekasi Trash Free'*. The program introduced the concept of community-based development, with the implementation of a Zero Government Budget to achieve the goal (Green Indonesia, 2011).

Bekasi entered into an emission-reduction purchase agreement with both environmental and social benefits, the first in Java. As part of the agreement, a gas-flaring facility was built by PT Gikoko Kogyo Indonesia, an engineering company that specializes in clean air technology, at the Sumur Batu landfill. This flaring technique captures the methane gas produced by the solid waste. Methane gas is a highly harmful gas therefore capturing this gas and utilizing it contributes to reducing the climate pollutants from this waste dump. The flaring facility will go a long way to reducing the number of climate pollutants deriving from Bekasi. As part of this program, the World Bank, acting as a trustee of the Netherlands Clean Development Mechanism, will purchase 250,000 tons of CO_2 Certified Emissions Reductions (CERs) per year for the next 15 years from PT Gikoko. These CERs are equal to one ton of CO_2 priced between US$5 to $10 (World Bank, 2012). The funds generated from this will be managed by the local community, along with the city government and PT Gikoko. This ensures that the local community is part of the decision-making process and also that they are jointly accountable for the use of these funds, with 7% of the revenue supporting community development and 10% added to the local budget. The new technology is expected to increase the supply of electricity to the landfill operation itself and also to the surrounding region. In addition, the power plant will help to reduce waste and pollution. The electric power generated from waste utilization is managed by PT Godang Tuajaya and initially produces 8 MW with a total production of 19 MW. It is estimated that in 2012, the power plant waste will be able to generate 26 MW of electricity. The electricity will then be sold to the State Electric Company and distributed to 50,000 houses.

Another solution proclaimed recently by the government to overcome the problems of waste was to turn the landfill into a landscaped park as a tourist destination. The idea may sound a bit ridiculous but as it turns out trash can be structured in such a way as to form a hill of garbage, then the garbage hill can be planted with grass, flowers and trees, creating a landscaped hill (Detik News, 2012). Furthermore, the park would be able to house educational facilities where people could learn about waste management and its utilization as a new source of material and energy. With the commitment of the government, it is not impossible that in a few years Indonesia will be the 'State with a Trash Tour Package'.

7.9. El Nido Resort, the Philippines

El Nido Resort is globally known for its commitment to sustainability through the 'responsible tourism' concept. Their commitment to the environment covers various programs such as marine turtle conservation, coastal and underwater clean-ups, nature-based guest activities, sustainable menus and providing environment education programs for staff and guests. Furthermore, they have skill training programs and try to foster local job creation. They also have an eco-efficiency program.

The El Nido Resort is located on the resource-protected area in the province of Palawan, the Philippines. It is 420 km from Manila and 238 km from Puerto Princesa, Palawan's capital. The El Nido Resort Group, commercially known as Ten Knots Development Corporation, operates two resorts: Miniloc Island Resort (1982) and Lagen Island Resort (1998). In 1984, the Ministry of Natural Resources established a 360 km^2 maritime area in El Nido as a turtle sanctuary. In 1991, the Philippine government proclaimed the rest of the bay area as a marine reserve, and in 1998 the Department of Environment and Natural Resources indicated the area as a 'Resource Protected Area'.

The El Nido Resort tries to address sustainability issues, including efforts towards eco-efficiency and zero-waste management. The nature of resort development is private with the resorts located in secluded sites, it is quite reasonable for El Nido Resort to have well managed waste management. One of the features to treat local waste is the Materials Recovery Facilities for the segregation and composting of solid waste. At the facility, the waste is separated into non-biodegradable and biodegradable waste. Biodegradable waste is composted into fertilizer and used in the landscaping of the resort as well as the Ten Knots farm. The non-biodegradable waste is sorted to separate that which can be recycled or reused and that which goes to the El Nido landfill.

The Materials Recovery Facilities (MRFs) at the El Nido Resort is a good example of eco-efficient development. Although the context of the project is not the urban area, the model of Zero-waste Management can be implemented in an urban context. For example, one urban neighborhood can have dedicated MRFs to sort and process local waste. The materials that are recovered can be used within the neighborhood (e.g. fertilizers to be used in landscaping and parks or recycled plastic to be used for kids' playgrounds) or sent to somewhere else as raw materials (e.g. recycled paper for tissue material).

7.10. Manav Sadhna Activity Center, Ahmedabad, India

Ahmedabad is the largest city in Gujarat, India. It is the seventh largest city and seventh largest metropolitan area of India, and has a city population of approximately 3.96 million people and metropolitan population of 5.56 million people. Ahmedabad is the fastest growing city of India, and ranked third as the fastest growing city worldwide. It is located on the banks of the River Sabarmati, 32 km from the state capital of Gandhinagar.

A major problem in Indian urban areas is the proliferation of slum developments and the urban poor. The slum settlements are primarily spread out around important railway stations, rail tracks (in the case of Mumbai) or around major roads. Mostly such settlements are in very important urban locations (for example the Dharavi settlement in Mumbai). These settlements embarrass the politicians and the elites in the cities, with the solutions proposed often aimed at just getting rid of the slums by putting the poor into high-rise, low-quality tower blocks or forcing them out of the prime properties. However, there is a way to co-exist with slum settlements. Authorities need to focus on developing the educational and ecological opportunities of the masses. An example of this type of solution has been recently shown by Ar. Yatin Pandya who designed an activity center in Ahmedabad.

The activity center at Rama Pir Tekra, Wadaj, in Ahmedabad has been one small attempt in the direction of recycling municipal (domestic) waste into building materials and hence is an eco-efficient city demonstration project. The activity center is located in the largest squatter settlement of Ahmedabad, and was created under the initiative of the social non-governmental organization (NGO), Manav Sadhna. The activity center is designed to be multi-purpose, serving as an informal school for young children, providing education for adults and serving as a training center and activity workshop for the creation of craft-based products created by women and elderly. The center also includes a dormitory, an administrative facility and a meditation facility for all religions. The campus was built using recycled municipal waste (World Architecture Community, 2012).

The Manav Sadhna activity center is an epitome of integrated architectural design which attempts to resolve several perennial problems of the developing world. Firstly, by recycling waste, the center helps to address issues of environmental pollution as such waste would otherwise simply get thrown away at dumpfill sites. Secondly, by converting the waste into building products, the center creates employment opportunities for the urban poor and thereby contributes to economic empowerment. Thirdly, the products utilized were locally developed, cheaper and better than the conventional alternatives, thereby making the building construction affordable. As an activity center, it functions truly for the poorest community of the city. The center is located in one of the largest slums of the city, which is inhabited by over a hundred thousand people, and gets used as an informal school for young children, a place providing vocational training to youths and in the evening, it turns into a community center with health facilities, recreational and religious activities and festivities. The center also holds medical camps for the settlement.

The center was built with the involvement of local unskilled and semi-skilled people and its construction techniques and detailing has been done keeping such a mode of production in mind. The building components were all developed through waste recycling and have been tested for their performance. Many of these materials have been found to be superior to conventional practices. The construction of the facility was applied integrally such that construction engineering and functional needs are aligned with aesthetic considerations, with these concepts emerging as mutually complementary. Wastes such as fly ash, dumpfill site residue, plastic bottles, glass bottles, crate wood, packaging wrappers and oil tin containers, amongst others, have been recycled as building materials (Figure 7.5).

The building has proven extremely cost-effective and has become an open catalogue for different walling, roofing, flooring and paneling options. The community along with other organizations have begun to emulate these construction options in their homes and so several other structures have been built. Thus, the center has been an effective role model to prove that cost is not a constraint but is rather a creative challenge, and as such, there is much creative opportunity in generating architecture for the have-nots. Contextually-appropriate architecture truly evolves from the needs and resources of its people and place.

The case study shows how social and economic issues can be solved together by integrating ecological design. By reusing waste such as glass bottles, wooden crates and construction waste as a building material,

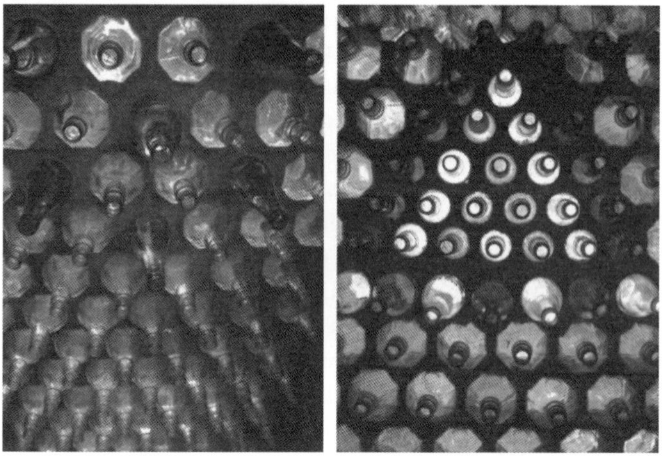

Figure 7.5 Walls Constructed from Waste Glass Bottles.

Source: World Architecture Community, 2012.

building costs have been significantly reduced and at the same time, the waste is also reduced. The activity center has had a great impact on the poor society by providing various programs, such as education, facilitating empowerment and community development. Eco-efficiency can be just a technical exercise or it can have a human face such as in Manav Sadhna.

7.11. Trashion, Indonesia

Indonesia is ranked the fourth most populous country in the world. It was home to 243,000,000 people at the 2010 census (Index Mundi, 2011). The enormous population is a sign of opportunity and also a threat in creating a sustainable environment in the future. While there is no significant breakthrough in large-scale waste management in Indonesia, the efforts to create a sustainable environment have occurred at the small-scale and community level.

The story of 'trashion' could be found in various cities in Indonesia. The term 'trashion' itself comes from the words 'trash' and 'fashion' and refers to the effort to convert trash, mainly plastic waste, into a fashion product. Trashion has mushroomed in the past decade with the phenomenon traced back to three situations that happened in the Indonesian community. The first situation was that many people

Figure 7.6 Trashion Products.

Source: Djumena, 2010a.

already had the basic skills necessary to create a fashion product, particularly sewing and pattern-making. The second is that Indonesia has an enormous amount of plastic waste, which is not being recycled or reused. The last situation is the existing awareness of the people about the environment. Trashion was encouraged through various initiatives including a guidebook, television shows and basic training from governments and other organizations. Some examples of trashion are shown in Figure 7.6.

Mrs. Ummah Daeng Ne'nang in Makassar started creating trashion after she was retrenched from her previous job. Ummah maintains that she was inspired to create trashion after watching a television show, and she continues to explore designs from a book called '*From Trash to Trashion: 25 Kreasi Limbah Plastik*'. The South Sulawesi government helped her by donating five sewing machines to help create a production line. Ummah could earn as much as Rp 3,000,000 per month, a much better income compared to the South Sulawesi minimum wage, which is only Rp 975,000. Whilst helping to create a better environment by recycling plastic waste, Ummah also helps to empower the surrounding community by employing housewives and creating new jobs for the unemployed (Djumena, 2010a).

In the city of Balikpapan, Mr. Achmad Iskandar got his inspiration from the 'National Environment Exhibition Day', which was held back in 2008. Currently, he produces 900 flower vases per month created from 600 kg of plastic waste. Achmad uses traditional motifs and patterns for his flower vase design. He also helps to empower the neighborhood and the community by creating more jobs (Djumena, 2010b).

Mrs. Clara Seiffie produces a product called 'CS bag' from plastic waste. She has 1,200 staff involved in her workshop, spread over four different cities. USAID helped support her production. The capacity of her production could reach as high as 10,000 items per month. She also exports her product to Italy (Kelola Sampah, 2010).

Mrs. Sri Wahyuni started her business in the trashion industry with help from the local government environment agency. They provided the basic training needed to recycle waste and, along with ten other housewives, Sri created 'The Waste Recycle Community' in Magelang and continues to produce trashion products. The demand for trashion comes from various cities particularly Jakarta and Bali which are major tourist towns. The Waste Recycle Community has managed to clean their village, empower the community and create more jobs and additional income (Entrepreneurship, 2010).

In Jakarta, Mrs. Heryanti Simarmata started her career in the trashion industry after winning a solid waste competition held by the Jakarta Green and Clean Program in 2008. She involved housewives from her surrounding neighborhood and with capital donated from the Unilever Foundation, Heryanti managed to acquire five sewing machines and employ ten staff to help her. Heryanti could earn as high as Rp 10,000,000 per month from her trashion products. She also expanded into the overseas market looking at countries like Singapore, India and UK (Kompas, 2010b).

The last story of trashion comes from Tangerang. Due to Indonesia's financial crisis in 1998, Mr. Slamet Riyadi was retrenched from his former company. Being unemployed, he started to see plastic waste as a resource material and started exploring the possibility of creating trashion products from it. Currently he has 70 staff, consisting of the elderly, trash scavengers and housewives. Mr. Slamet has successfully exported his product to various countries including Brunei, Singapore, Egypt, Nigeria, Holland, Italy, Canada and Australia. He is also very active at campaigning for environmental awareness, visiting schools in Java (Kompas, 2010a). He believes that the majority of the people and the community already have knowledge and awareness about sustainability. A catalyst, such as a workshop or a guidebook is all that is needed to turn this potential of creating fashion objects from trash into a real act to create a more eco-efficient city, whilst contributing to the economic and social aspects of the city.

7.12. Conclusion

Eco-efficiency concepts provide a model for how green urbanism in Asia can be implemented at national, regional and local scales. Singapore provides an example of how a nation, initiated by government, could apply a comprehensive ecological approach to manage waste and water. At the regional scale, cases from Japan, China, Korea, Thailand and Indonesia were discussed, looking at those that balanced contributions between the public and private sectors. The local scale examples provided a more bottom-up approach where usually the local governments were not involved. The case studies from Indonesia, India and the Philippines were presented to show how the movement originated, was sustained and how it then progressed further in order to achieve the goal of an eco-efficient city.

Asian cities are dense and rapidly industrializing hence the waste and pollution from the metabolism of materials can be highly significant. Only by applying eco-efficiency concepts can these cities manage to be liveable and productive. The evidence provided in this chapter shows that the concepts are being employed and do work in Asian cities, both in wealthy cities of Singapore, Japan and South Korea as well as the new cities of China and the less wealthy cities of Asia. The rapid spread of this aspect of green urbanism is one of the hopeful trends in Asia.

CHAPTER EIGHT

The Place-based City

8.1. Introduction

Imagine a city where people's daily lives are conducted around nature and water, where the buildings seem to grow from a soft, green landscape, a city where the natural environment holds great meaning to the residents and where residents feel responsible for the well-being of the environment. The city is connected to the natural world and this is reflected in the everyday built environment of the city. These are the characteristics of a place-based city. These are the characteristics of a city with a strong sense of place — where water in particular plays an important part in the culture of the city.

Sense of place is where people recognize, enjoy and celebrate the unique qualities of a place, including the built and natural environment (Newman and Jennings, 2008). It is also what makes places different from each other. More than this, sense of place is the emotional connection people have to the everyday locations of a place and the physical manifestations of that emotional attachment. Such feelings for a place enable strong local communities (Beatley, 2004).

Enhancing a sense of place brings out the human dimension of green urbanism and plays a crucial part in building resilient cities. It is no longer uncommon for cities and regions to develop their local economies and community pride by fostering a unique and special sense of place. Developing a sense of place creates social capital from which a community can benefit. Newman *et al.* (2009, p. 82) explain that "when people belong and have an identity in their town or city they want to put down their roots and create local enterprise". A tight-knit community often results in demand and support for local businesses.

The strength of community integration helps generate respect and care for the environment, responsible management of resources and acceptance of new green technologies.

Newman and Jennings (2008, p. 144–155) developed five strategies for cities and communities to enable them to foster a sense of place. These are:

1. Protect important existing elements of natural and cultural heritage;
2. Design to make historical and current social and ecological processes more visible;
3. Connect the urban form to the wider bioregion;
4. Use cultural practices and the arts to nurture and deepen a sense of place; and
5. Discover a city's 'songlines', i.e. indigenous stories.

These strategies are all complementary and are demonstrated by the case studies presented in this chapter. Furthermore, these strategies highlight that both natural elements, particularly water, and cultural elements, art and design, are important in enhancing sense of place. Four overarching themes have been identified, specific to the place-based city examples from the Asian region. These themes are:

1. Water-sensitive urban design;
2. Reinterpreting traditional forms of the architectural vernacular;
3. Participation from the community; and
4. Public art.

These themes form the structure of this chapter.

8.2. Water-sensitive Urban Design

Many cities have been actively reclaiming and transforming their urban waterways from engineered drains back to natural water systems. Sacred places are usually associated with water and with access to water. In Asian cultures especially, water has always been seen as sacred and as a source of spiritual purity, playing an important part in the traditional building and planning sciences, such as the Chinese *Feng Shui* (literally 'Wind Water') and the Korean *Pungsu-jiri*, the study of the patterns of wind and water. Hence, the reclamation and regeneration of waterways, rivers, creeks and waterfronts is a way that communities can reconnect with their natural environment.

The South East Queensland (SEQ) Healthy Waterways Partnership defines Water-sensitive Urban Design (WSUD) as the "approach to the planning and design of urban environments that supports healthy ecosystems, lifestyles and livelihoods through smart management of all our waters" (South East Queensland Healthy Waterways Partnership, 2010). This essentially means that city development is sympathetic to the natural flows of water and that water management practices follow natural processes. The first case study looks at how Singapore has managed its urban waterways and water storage. The second case study describes restoration of a river in Seoul, South Korea, that was buried beneath a freeway. The third case study presents the rehabilitation of the Pasig River in Manila.

8.2.1. *Singapore's Urban Waterways*

Water is the life blood of many Asian civilisations, particularly in the coastal regions of Southeast Asia, where Singapore is located. Access to water has historically enabled trade and travel. In Singapore, since the time of Sir Stamford Raffles, when trade was conducted on the banks of the Singapore River, water has always been associated with trading and with economic prosperity. The environmental identity of residents in Singapore is connected to with the waterways and parks. This is reflected through cultural icons such as the Singapore riverfront. The redevelopment of Marina Bay and the Bedok Reservoir exemplify this connection to water.

8.2.1.1. *Marina Bay Waterfront*

Given Singapore's long relationship with water, it is not surprising that much of the city's iconic architecture has been constructed along waterfronts as a symbol of pride and success. Marina Bay is home to many iconic buildings including the Esplanade theaters and the more recent Marina Bay Sands Integrated Resorts (MBS IR) (Figure 8.1). At the opening of the Marina Bay Financial Center in 2006, Singapore's Minister for National Development, Mr. Mah Bow Tan, announced that Marina Bay was envisioned to be a "seamless extension of the Central Business District, humming with business and activity, and sizzling with entertainment 24/7" (Mah, 2006). The vision for Marina Bay was to be not just part of the commercial and economic center of Singapore, but also as a "live, work, [and] play environment" for people (Koh-Lim, n.d).

Figure 8.1 Iconic Architecture at Marina Bay.

Source: Peter Newman.

This vision of a community within the city center resonates well with the idea of 'place-basedness'. Not so long ago, the central business area was deserted after business hours and on weekends. The Marina Bay development was seen as a way to bring life back into the commercial center, and meant that people, if they wanted, could live close to their place of employment. Furthermore, the area was designed to host community, national and international events and to provide a place for civic pride.

In 2008, the Marina Barrage was built, closing off the mouth of the Singapore River, and turning Marina Bay and Kallang Basin into a freshwater reservoir, the Marina Reservoir. The area now provides a supply of freshwater and recreational areas and helps to control flooding. Along with open spaces along the reservoir, the barrage was built with a sloping green roof providing open spaces accessible to the public and has become a popular destination for picnicking and for flying kites.

As discussed in Chapter 5, much of the development in Marina Bay utilizes and celebrates green technologies. Along with fostering a sense of place and community, the green roof at the Marina Barrage is also environmentally sustainable in that it reduces the surface heat of the building. The barrage also is home to a solar research facility. In 2009, the Marina Bay Financial Center received a number of Gold 'Green Mark for Buildings' awards from the Building and Construction Authority (BCA). These awards recognize building owners and property developers who construct and maintain environment-friendly buildings. One feature of the iconic Marina Bay development is the use of cold water pipes set in the deep water of the reservoir to help with air conditioning in the building (see Chapter 5).

The Marina Bay development is built around the concept of 'working in a garden'. Through sky terraces, linear parks, green open spaces and shade-covered pavements; Marina Bay provides a walkable environment. It also serves as a backdrop of many local television series, and is a familiar symbol to all Singaporeans. Marina Bay's floating platform is used as the location for the nation's National Day Parade, illustrating the links between place and national unity.

The Marina Bay development, although still evolving, has been quite successful in enabling a place-based community in the heart of Singapore's city center, created from almost nothing. The development has brought vibrancy to the area and has demonstrated how to turn a waterfront and water management area into a multi-purpose place with iconic qualities.

8.2.1.2. *Bedok Reservoir*

Another initiative by the Government of Singapore of using WSUD for water management and to enhance sense of place is the creation of a reservoir park, the Bedok Reservoir Park, in the residential neighborhood of Bedok. This initiative is part of a city-wide scheme to revitalize urban waterways within Singapore. Bedok Reservoir Park was created in a former sand quarry and has proved to be a popular park. The park is strategically located and easily accessible to a number of schools, residential housing estates and healthcare facilities, allowing for it to be

Figure 8.2 Revived Waterway in Singapore Surrounded by Housing, a Hospital and Public Space.

Source: Peter Newman.

utilized by all the community. The park enables biodiversity and brings nature closer to people (National Parks, 2009).

In the past, especially during the 1960s, the waterways in Singapore were polluted. The government has been actively cleaning up the polluted waterways and trying to create healthier environments where water can be part of everyday living (Figure 8.2). The Active, Beautiful, Clean (ABC) waters program is an initiative which aims to enhance water amenity by going beyond the functional uses of freshwater reservoirs (Ministry of the Environment and Water Resources, 2011). Instead of restricting public access to the reservoirs as was the management practice in the past, the new strategy is to enable public access to the water and to promote social responsibility for the waterways and public spaces. In doing so, this strategy enables a sense of ownership of the waterways. Signs erected around the Bedok Reservoir promote the ABC water initiatives and inform the public about cleaning efforts. Boardwalks located around the reservoir increase accessibility to the waters. Major activities have been staged within the park to promote it as a space for the community, ranging in scale from the large Chingay River Festival to smaller events such as school excursions.

A 3.8 km green corridor is being created along the reservoir to connect the Bedok Reservoir Park to the popular East Coast Park. The corridor will be lined with community gardens, recreational points and sports facilities, such as exercise equipment that would be suitable for people of all ages. In the middle of the corridor is the Bedok Town Center, with the Bedok Mass Rapid Transport (MRT) station and the bus interchange. The idea is to bring the outdoors closer to residents in a pedestrian-friendly environment.

Bedok Reservoir Park is an interesting example of how the regeneration of waterways can serve functional purposes of water provision and simultaneously promote social interaction and a sense of place.

8.2.2. *Cheonggyecheon Restoration Project, Seoul, South Korea*

Cheonggyecheon, or 'clear valley stream', is a 5.8 km stream flowing through downtown Seoul in South Korea. During the Joseon Dynasty (1392–1897), Cheonggyecheon was one of several streams that drained from the surrounding ring of mountains into the Han River. The river, as depicted along Cheonggyecheon through artworks, was a vital part of the residents' lives. However, even historically, Cheonggyecheon and the Han River were also seen as a place for wastewater disposal. By the 1950s, after the Korean War, the river had become a wide, shallow stream where water could not flow naturally and was a drain for the surrounding slum areas. It had become an unhygienic stretch of wasteland. The slums were cleared and the stream was paved over to form the Cheonggye Road. In 1968, the construction of a 16 m wide, 5.6 km long elevated highway, the Cheonggye Expressway, began (Figure 8.3). This expressway was seen as a symbol of progress in the industrializing economy.

As Seoul expanded, southern Seoul, Gangnam, was developed into a prospering new downtown while the old Central Business District (CBD) in Gangbuk, north of Seoul, declined. Increasingly, there was a need to balance out the two sides of the city. Furthermore, after Korea's

Figure 8.3 Cheonggye Expressway, 1976.

Source: gardenvisit.com.

economy stabilized, the focus of the nation shifted to environmental concerns. The expressway became seen to be contributing to noise and air pollution in the Seoul downtown (Kelley, 2008). In early 2000, a plan was devised for Gangbuk to be revived as an eco-friendly downtown where history, culture, business, commerce and tourism could come together. One of the key aspect of this revitalization was the restoration of the Cheonggyecheon River and the removal of the expressway that covered it. The idea was that the greening of Seoul would transform its image from one of 'grey concrete' to one of 'a lush, green city where clear waters flow'. Seoul wanted to be re-branded as a 'human-oriented' and 'environment-friendly' city (see also Chapter 6). In practical terms, the dismantling of the highway was inevitable as the infrastructure was assessed to be past its life expectancies and was no longer considered safe. Uncovering the Cheonggyecheon was taken to be able to facilitate these aims, particularly enabling the revitalization of the surrounding neighborhood (Park, n.d.).

Under the leadership of the then Mayor Lee Myung-Bak (now the president of South Korea), the Cheonggyecheon Restoration Project was completed in just two years. To do this, an integrated approach was adopted involving the city's government, a citizens' committee and a research team. To convince the public of the feasibility of the project, the project leaders had to allay fears regarding traffic congestion from the removal of the highway. The citizens' committee suggested that the research team conduct live transport simulations. Residents also lobbied for people to reduce car usage during the period of construction. Seoul's City Council responded by setting up pedestrian way-finding signage to increase the walkability of the streets (Sustainable Cities, 2011). Public participation was enabled through surveys conducted by the citizens' committee. These surveys gave citizenry a voice in the decision-making process during the development of the project. The restoration of the river was completed in 2005.

The project was tied to the political pledge of Lee's mayoral campaign and then to his campaign for presidency. This did cast some negative light on how locals viewed the Cheonggyecheon project (Kelley, 2008). However, despite some criticisms, the river restoration has proved to be extremely popular, becoming a place where both locals and tourists flock to. Weekly, approximately 500,000 people walk along the river promenade (Kelley, 2008). Informal seating areas along the banks of the stream provide space for families and children to take a breather, dip their feet into the cooling waters and enjoy the vista (Figures 8.4 and 8.5).

The popularity of the river is partly due to some quantifiable benefits gained from the river's restoration. These benefits include improved air and water quality, decreased noise levels and a decrease in the urban heat island effect. The river restoration has resulted in an average temperature drop of 3.6°C in places 400 m around the river. The number of vehicles entering the downtown area was reduced by 2.3%. This corresponds to an increase in the number of people using public transport, with bus use increasing 1.4% and subway use 4.3%. These modes carry an estimated 430,000 people daily. A biodiversity count in 2009 revealed that 213 species of plants, fish, birds and other organisms are thriving in the restored habitat (Stein, 2009). The property values along the stream also increased (Biggs, 2010) (see Chapter 6 and 7 for some of the restoration details).

One of the aims of the Cheonggyecheon restoration was to enable the city to "rediscover its historical roots"(Park, n.d., p. 11). Artefacts found during the excavation of the concrete paving, such as foundation stones of the old bridges, were kept and displayed along the river. These artefacts validate and reinforce the sense of pride the people

Figure 8.4 The Start of the Cheonggyecheon River Promenade.

Source: Peter Newman.

have in their 600-year long history. An ancient ceramic Korean narrative is displayed along parts of the river promenade (Figure 8.6). There are also small place-markers along the stream that illustrate its history. One example of these markers is the 'Laundry Site', which recalls a time when the river was used as a site for conducting daily activities like washing clothes (Seoul Metropolitan Government, 2011). The Cheonggyecheon Culture Center, a six-story building along the stream, tells the story of the river's past and present, the city's 'songline'. Sandercock (as cited in Newman and Jennings, 2008, p. 154) argues that "the sharing of stories can bring people together and help them shape new collective stories". The setting up of the Culture Wall on the upper stream of *Ogansumun* (floodgates) provided opportunities for local history and culture to be flaunted. The project engaged five contemporary artists to explore the theme of 'Harmony between Man and Nature' (Seoul Metropolitan Government, 2011). Their artworks, which display a positive message for environmental awareness, are

Figure 8.5 Cheonggyecheon River Promenade.

Source: Peter Newman.

appreciated by visitors. The project invited 20,000 citizens to decorate 10 by 10 cm tiles to be displayed as a 'Wall of Hopes' (Seoul Metropolitan Government, 2011). This wall creates a feeling of closeness between people and the built environment as it gives them a small stake in creating part of it. The many plazas built into the plan provide ample space for local and international artists to perform. These different forms of creative expression help to celebrate, and subsequently, deepen the sense of place.

The restoration of the river has not all been a perfect picture however. The construction of the Cheonggye Expressway in the 1970s attracted many merchants on both sides of the road. The removal of the expressway meant these merchants had to be relocated. The intent and effort in setting up of a citizens' committee was to try to resolve the conflict between these merchants and the government. There has been criticism of the project as gentrification of the area. The project is also not sustainable in the fact that it requires energy-intensive pumping of

Figure 8.6 People Enjoying the Cheonggyecheon River Promenade and River Stone Crossing. An Ancient Ceramic Korean Narrative is Displayed along Parts of the River Promenade.

Source: Peter Newman.

water from the Han River. Many people, however, consider the benefits of the restoration to outweigh these adverse consequences (Vidal, 2006), with many of these benefits not being quantifiable.

Despite the inadequacies highlighted, the demolition of the viaduct for the restoration of the stream is a radical move that sends a strong message. A number of the freeway columns have been left in the Jonchigyogak area of the stream to remind people of the evolving history of the area. These remind people about the possibility of change and the evolution towards a green, place-based future. The restoration of this river has been an exemplary project which other governments, such as those from Shanghai and Tokyo, use as an example of what is possible (Vidal, 2006). The Cheonggyecheon provides a symbol of the new green economy and is a symbol of place in green urbanism.

8.2.3. *Pasig River, The Philippines*

Pasig River, located in The Philippines, is a 27 km river that connects Laguna and Manila Bay, and serves as an important transportation route. Four major river tributaries and 47 creeks and esteros (estuaries) drain into the Pasig River. Throughout the country's history the river was the city's lifeline and center of economic activity. With industrialization, factories, including a major oil depot, were developed along the river and the river became a place to dispose of liquid wastes. Manila has also experienced rapid population growth, growing from 219,928 in 1903 to 11.5 million registered residents in 2007. This population is predicted to be closer to 16 million if you include non-registered residents. Pollution in the Pasig River was compounded by garbage entering it from its tributaries and informal settlements constructed along the river's edge (Philippine Green Movement, 2009). The pollution of the river became hazardous and drove inhabitants and visitors away (Figure 8.7). The river was declared biologically dead in 1990.

The river needed rehabilitation. The Pasig River Rehabilitation Commission (PRRC) was established in 1999 to ensure that the Pasig River was returned to its historical state and able to be used for transport, recreation and tourism. They were given four years to complete the rehabilitation in phase (Pasig River Rehabilitation Commission, 2012). The project became top priority for the government. Sub-organizations were set up to oversee different tasks, for example overall project

Figure 8.7 Pollution of Pasig River.

Source: istockphoto.com.

Figure 8.8 The Pasig River.

Source: Luis Liwang, retrieved from Multiply, 2009.

management, managing solid and liquid wastes and improving living conditions of residents (Department of Environment and Natural Resources, 2011). Figure 8.8 provides a view of the rehabilitated river.

More than a hundred groups from the public and private sectors joined in the rehabilitation efforts. One program was a community volunteer group called 'River Warriors' (Galarpe, 2010). River Warriors undergo training in law enforcement, leadership, discipline, personal development and self-defense, along with environmental education, and take an oath to safeguard the Pasig River (Figure 8.9). Leadership was the most important component of this program (Galarpe, 2010). The River Warriors have been key in the rehabilitation of the river. Another example of an organization involved in the rehabilitation of the Pasig River is the Clean and Green Foundation. The foundation sources funds from individuals and the private sector to be given as awards to residents who keep their section of the river the cleanest.

Rivers are vital to the sense of place. They provide breathing spaces within cities. Water is an important aspect of this, with financial and psychological benefits.

Figure 8.9 'Knighting' of Pasig River Warriors.

Source: Galarpe, 2010.

8.3. Reinterpreting Traditions and Vernacular within Natural and Cultural Heritage

Protecting and restoring key elements of a city's natural and cultural heritage is one of the five strategies suggested by Newman and Jennings (2008) in order to strengthen the sense of place. This requires a continuation of traditions and culture. The problem is in keeping traditions relevant to current contexts in order to stop reinterpretations from becoming empty imitations of the past. Japanese builders have figured out a way to do this, enabling them to pass down their traditional building techniques without the techniques becoming obsolete. The Shinto Ise Grand Shrine is rebuilt every two decades. Each new iteration utilizes traditional techniques, along with relevant newer technologies and materials. This ensures that the old construction vernacular is kept current. While the Ise Shrine precedent provides an example at the scale of an architectural object, the Urban Tulou in Guangzhou, China, provides an example of a reinterpretation of traditional vernacular at a community scale.

8.3.1. *Urban Tulou, Guangzhou, China*

Traditional architecture often possesses environmentally-sensitive qualities and passive design strategies which can inform and inspire

Figure 8.10 Traditional Hakka Tulou.
Source: Saieh, 2009 (Photographer: Iwan Baan).

current architectural design. The Urban Tulou project in Guangzhou, China, provides an interesting example of a reinterpretation of traditional architecture vernacular in a modern context (Saieh, 2009). The Tulou is a traditional Hakka housing typology, generally found in the southern parts of China where there are hot summer conditions similar to Guangzhou's (Figure 8.10). The Tulou features living areas that face an internal courtyard space, providing shaded outdoor areas. Windows (or openings) that face the external space are smaller, allowing for natural ventilation while still protecting the internal areas from the summer sun. The Tulou typology has been reinterpreted by the Chinese architectural firm, Urbanus Architects, for a 220-unit apartment complex for a low-income migrant housing project in Guangzhou (Saieh, 2009) (Figure 8.11).

The Urbanus Architects' reinterpretation has a mix of dwelling sizes along with ample public spaces including an internal courtyard (Murray, 2010). The internal courtyard design allows access to sunlight and shade as needed and is an important part of enabling a community atmosphere. This typology has provided open public spaces for community living to offset the small living units. The

Figure 8.11 Urbanus Architects' Reinterpretation in Guangzhou.
Source: Saieh, 2009 (Photographer: Iwan Baan).

housing unit allows for safety through passive surveillance by providing visual connectivity, creating an invisible network of trust in the community.

The reinterpretation of the Tulou housing in Guangzhou has been largely successful as a design solution for modern day housing concerns compared to most migrant housing in Guangzhou. This innovative reinterpretation has successfully allowed daylight and natural ventilation into the spaces, reducing electricity requirements and is a good model of how to design sustainably and cost-effectively for the future as well as using design to build on traditions of place. The project was shortlisted for the Aga Khan Award for Architecture in 2010.

8.4. Sense of Place, Pride and Participation

This section provides case studies of community empowerment. These case studies demonstrate a process of community building in low-income areas that enable people to obtain new knowledge and to improve their existing knowledge regarding their community and the future of their community.

8.4.1. *One Village One Playground, Babakan Asih, Indonesia*

The One Village One Playground project, initiated by Urbane Indonesia, involved the creation of a play space for children in a slum area in the village of Babakan Asih, Bojongloa Kaler district (formerly Blok Tempe), South Bandung, Indonesia. The village is geographically ideal with numerous services, such as markets, a central city square, a central bus terminal, a recreation center and a hospital all located within easy walking distances; however, the area lacked open spaces, particularly healthy and safe spaces for children (Urbane Community, 2010, p. 1). The vision was to enhance the quality of the urban environment, to improve the existing conditions and to provide play space for the local children.

The main objectives of One Village One Playground were to:

1. Create public spaces that are safe for kids to play and grow in.
2. Build public awareness to create a healthy environment.
3. Educate the local community to create independence for managing the environment.
4. Use public participation to support the empowerment efforts for an economic community.

Part of the One Village One Playground project was also to provide good quality urban environments; therefore the project also included activities such as infrastructure improvements and flood control measures. The environmental preservation program part of the One Village One Playground project furthered a 2008 community-initiated development project called 'Sustainable Bandung' that was a collaboration with the Common Room Networks Foundation (Common Room Network Foundation, 2011). At the core of this program were regular discussions and brainstorming sessions aimed at recognizing and developing alternative solutions to improving the ecological conditions of Bandung. As part of this, local residents developed maps that identified the topographical contours of the village to enable a new drainage system to be designed for the flood-prone area.

The One Village One Playground project was funded by private developers and it adopted a multidisciplinary approach that revolved around collaboration between the local community and professionals in the field of design, planning and architecture. Through a series of brainstorming exercises, rigorous research and design critiques, the public played an invaluable role in determining the final outcome of the project. This sense of responsibility and ownership also helped in creating a sense of place and belonging.

A more practical concern was the provision and ongoing maintenance of 17 water wells. These wells are used to minimize the frequency of flooding during the rainy seasons and allow rainwater to be stored for use. The community's role in the building of the wells and the visible improvements which resulted from the project, facilitated a sense of pride in the local communities and allowed them to be more proactive in building their environments for the better. Since the system was put into place, flooding is now reduced to only half an hour as compared to floods lasting seven days before the construction of the 17 wells (Figure 8.12).

With the new source of water, the local community in Babakan Asih have taken to growing catfish in the wells during the rainy seasons. This functions to 'clean' the well of mosquito larva and minimize the spread of malaria and dengue fever. The fish are also bred for sale in the community market nearby. The presence of the well has revitalized the use of the public space: it is now bustling with activities (Figure 8.13). This project has been an inspiration to many other communities across Indonesia, with more people understanding the benefits of creating

Figure 8.12 Local Communities Helping in Well Construction.

Source: Urbane Community, 2010.

Figure 8.13 New Public Space of Babakan Asih Village.

Source: Urbane Community, 2010.

sustainable habitats for living that improve the sense of place in a community.

8.4.2. *School-building: METI in Northwest Bangladesh and Green School in Bali*

This section describes the building of two schools, METI in northwest Bangladesh and Green School in Bali. These schools illustrate the importance of enabling a community to be involved in civic life and the importance of community empowerment as the basis for an enhanced ecological sense of place.

8.4.2.1. *METI — Handmade School*

The METI (Modern Education and Training Institute) School in Rudrapur, northwest Bangladesh (Figure 8.14), was an initiative by Dipshikha, a local NGO made up of social workers, teachers and youths, with the main aim of helping "people learn about the value of the village . . . [and instilling] in the children self-confidence and independence with the aim of strengthening their sense of identity" (Heringer, n.d.). It was hoped that by improving the quality of rural living it would help reduce the migration of villagers to the city. The architects, Anna Heringer and Eike Roswag, spearheaded the design and construction

Figure 8.14 The METI Handmade School.
Source: Krisenki, 2008.

of the school, engaging and working closely with local craftsmen, students, parents and teachers. This strong partnership enabled the school to be constructed in just four months. The school was built entirely by hand, hence its informal name: 'The Handmade School'. The school was the recipient of the Aga Khan Award for Architecture in its Tenth Cycle, 2005–2007 (Aga Khan Award for Architecture, 2010).

The school set out three strategies (METI School, 2011):

1. The construction of representative, public buildings to give a positive image to earth buildings.
2. The construction of residential buildings as a model for low-cost and environment-friendly housing.
3. The training of local craftsmen and constructors as a basis for sustainable implementation.

The master plan for the project was executed in two phases. The first phase saw the completion of the school's construction. The second phase involved the construction of residential dwellings for students.

The design of the school was based on the traditional earth buildings of Bangladesh (Heringer and Roswag, 2010). The problem with this way of building, however, was that it was not entirely waterproof and therefore, prone to quick deterioration. The architects devised a way of waterproofing the structure during the process of design. Additionally, they introduced a masonry foundation to reinforce the structural integrity of the traditionally fully-earthen structure. These tweaks to the existing technique made earth buildings relevant to the community once again. Twenty-five local tradesmen were trained on the job during the construction process (Heringer and Roswag, 2010).

The school was constructed out of bricks, bamboo and a mud and straw combination. Bricks are the most common building products of Bangladesh's manufacturing industry. These are made from the clay-based soil found locally. As there were no brick-makers in the village, the masonry construction was outsourced to a company in Dinajpur, the regional capital 20 km away. Local laborers in Rudrapur were assigned the earth works and bamboo construction. This is an example of not only encouraging local enterprise but also connecting to the larger bioregion to generate regional business due to a local demand (Heringer and Roswag, 2010).

Architect Anna Heringer describes that the construction of the project did not use cranes "but a lot of people... 400 tonnes of wet loam were only moved by human labour and cows/buffaloes. Instead of wheelbarrows the Bangladeshis balanced the load in baskets on top of their heads". Students at the school study the standard government syllabus in the morning and then take extra-curricular activities in the afternoon. The school construction was part of the extra-curricular activities. The empowerment of local people and their children through enabling them to sculpt their own living environment helped to instill a sense of pride in the community.

8.4.2.2. *Green School, Badung, Bali, Indonesia*

The Green School, located in Badung, Bali, focuses on providing a holistic education to its students. Saieh (2010) describes:

> *Environmentalists and designers John and Cynthia Hardy wanted to motivate communities to live sustainably. Part of that effort was to show people how to build with sustainable materials, namely bamboo. They established the Green School, and its affiliates: the Meranggi Foundation, which develops plantations of bamboo plants through presenting bamboo seedlings to local rice farmers; and PT Bambu, a design and construction company that promotes the use of bamboo as a primary building material, in an effort to avoid the further depletion of rainforests.*

Students come from all corners of the world, with local Balinese children funded by scholarships, allowing them to benefit from an international standard of education infused with Balinese traditions and cultures. The curriculum for younger children is influenced by the work of Rudolf Steiner, who pioneered the idea of holistic education. Older students have the opportunity to study many qualifications including Green Studies, which focuses on sustainability, and a Creative Arts program that complements the academic curriculum.

The sustainable campus features a range of architecturally significant spaces, including large multi-story communal gathering places, classrooms, a gym, faculty housing, offices and cafes. Bamboo, grown locally using sustainable methods, is used in innovative and experimental ways to demonstrate architectural possibilities (Figure 8.15). The result is a global green community with a strong mandate for education, which encourages students to be curious, engaged and passionate about the environment and the planet.

Osanai (2009) explains the approach of the Green School as one that:

> *... helps to create young adults with leadership skills in the spirit of the green movement, taking what they learn here and spreading their wings. They see the natural world in its intended state and then work in metropolises...with a unique perspective. The founders of Green School believe raising children to see the polar opposites of life prepares them to be active in the world at large. For this reason, they recruit students from across the globe. Although it's an international environment, English and Indonesian are used throughout the daily classroom work.*

Students stay with local families and see for themselves how the people of the island live their daily lives. These experiences enable the students to have a bigger picture of the world.

The projects run by the Green School aspire to enable the local community to understand the basic principles of creating a green and sustainable community and to enable a unique sense of identity and place. The process of involving children at a young age helps to foster a sense of social and environmental responsibility within them. Equipping the students with hands-on learning in the skills of construction promotes a practical and methodical approach to discreetly improving the environments in which they reside. The project also provides local children with

Figure 8.15 The School's Buildings are Mostly Made from Bamboo Used in Innovative Ways.

Source: Green School, 2012.

education, hopefully providing them with opportunities to escape the poverty cycle, which is also an important aspect in shaping the community for the better. Therefore, the project can be said to be highly successful at many levels.

8.5. Encouraging Public Art

Public art is often a less permanent and less extensive way of using cultural practices and the arts to nurture and deepen a sense of place (Newman and Jennings, 2008, p. 153). Its small scale creates a tendency for its usefulness to be overlooked, but it is done on a human scale so it can be a powerful tool to engage human emotions and impact on behaviour. Forms of public art range from installations at cultural festivals to architectural objects like pavilions or memorials in public parks and gardens. Increasingly, landscape architects are utilizing urban public spaces for art using greenery (Figure 8.16). Public art complements walkable areas, creating delightful experiences and surprises in urban environments.

Public art can also be a form of activism, generating awareness of the environment (Justa, 2009). The biophilic urbanism outlined in Chapter 6 is also an example of ecological public art with a strong sense of place.

Figure 8.16 Brophilic Art along the Cheonggyecheon River, Seoul.

Source: Peter Newman.

8.6. Conclusion

A city-wide green urbanism agenda in protecting natural and cultural place heritage would lead to an urban design program that makes visible these socio-historical and ecological processes. The case studies reviewed in this chapter involve public buildings and infrastructure that are used by the community on a daily basis. These highly visible landmark projects are demonstrations of hope. Singapore had a vision of being a 'City in a Garden', and despite land scarcity, has tried to keep the presence of the greens and blues of nature obvious amongst the concrete. Seoul unsentimentally ripped out a stretch of highway to restore a river that is now a hive of activity buzzing with the city's life. Manila rescued a river under threat of drowning in environmental waste, restoring it to its rightful place in the lives of its people. Readapting traditional building typology, as shown through the example of the Urban Tulou, was a method of extending a culture's legacy into modern contexts, while public art could add value in enriching a community's artistic and cultural heritage.

A bottom-up approach, while not completely without its loopholes, ensures a longer-lasting and more successful implementation of any green urbanism practice, especially in developing a sense of place. In areas where the people have yet to find strength as a community, usually in low-income regions, external forces (such as non-governmental organizations or architects with a humanitarian focus) need to facilitate localism by volunteering their help in empowering these communities to enable a better quality of life. Building up local employment decreases rural-urban migration. It also restores confidence in the communities' ability to be self-sufficient, as can be seen in the case studies of community participation in rural Asian villages.

When used responsibly and ethically, place-based programs can contribute to building a society's collective memory, which is needed to build green urbanism. This chapter illustrates how Asian cities are beginning to show this in a range of iconic projects.

CHAPTER NINE

The Sustainable Transport City

9.1. Introduction

Transport is the most fundamental infrastructure shaping a city. This chapter provides a discussion of cities in Asia that utilize sustainable transport options in order to facilitate the transition to green urbanism. The cities discussed include: the new metros in Delhi, India and Shanghai, China; bus rapid transit (BRT) and bike sharing in Guangzhou, China; a move towards more sustainable transport in Yogyakarta, Indonesia; a bike sharing program in Taiwan; the walking plaza in Pune, India; the BRT program called Janmarg, in Ahmedabad, India; and an 'E-jeepney' program in the Philippines.

9.2. Delhi: Transforming Itself to Achieve a Sustainable Transport System

Delhi, officially known as the National Capital Territory of Delhi (NCT), is the largest metropolis in India by area (1,483 km^2) and the second largest in terms of population (14 million people). Delhi is an old city that has developed around different transportation modes, from walking to bullock carts to cars. Delhi is continuing to transform itself through the implementation of a metro system and the use of compressed natural gas.

Delhi has nine districts and is easily identified into three main parts: Old Delhi, Colonial Delhi, and Other Delhi, which includes east, west, south Delhi. Colonial Delhi, designed by the British Government, is in the center and includes Connaught Place, Delhi's central business district along with the President House and India Gate. This is primarily a government and business area with little residential population. Old Delhi, or Shahjahanabad, is a compact walled city built around the transport modes of walking and bullock carts. The city was founded by Mughal Emperor Shahjahan in the 1600s and the area is home to many of the city's historical monuments such as the Red Fort and Jama Masjid. Of particular importance is the famous market, Chawri Bazaar. The roads in the area are narrow and crowded, filled with pedestrians, scooters and rickshaws.

Between 1981 and 1998, Delhi experienced a rapid rise in both population and car ownership. Congestion and air pollution became a major concern of the Delhi government. The concept of a mass rapid transit system first emerged from a traffic and travel study carried out in the city in 1969. Various studies were carried out looking at possible mass rapid transit options, and in 1984 the Delhi Development Authority and the Urban Arts Commission developed a proposal for a multimodal transport system in the city. The Delhi Metro Rail Corporation (DMRC) was established in 1995 and construction on a metro system started in 1998 with the first line, the Red Line, opening in 2002 (Delhi Metro Corporation, 2010). The network now has six lines, with a total length of 189.63 km and 142 stations. It is a mixture of elevated, surface (or at-grade) and underground lines and carries an average of 1.8 million people per day. The creation of the Delhi metro was the first step towards creating a sustainable transport city and the project has become the lifeline of the Delhi people (Figure 9.1) and a symbol of peace and harmony (Figure 9.2). The underground Chawri Bazaar station has become an important node for the community living in the area. The station has revived the area by connecting it to the rest of Delhi.

In 2010, the metro was extended to Gurgaon a new town in the southwest metropolitan area. Gurgaon is home to 1.5 million (in 2011) and is a major employment hub due to its close proximity to the Delhi International Airport. The area was linked to Delhi by an eight-lane expressway, which was extremely congested and averaged seven fatal road accidents a day. The linking of the area to the Delhi metro has been a catalyst for Gurgaon to enforce better traffic planning within the city.

Along with the metro line, the use of Compressed Natural Gas (CNG) to help combat air pollution in Delhi has been a remarkable story (Figure 9.3). Part of the reform came from the people protesting the air pollution created by the rise in private vehicles (Figure 9.4). In November 1996, the Supreme Court of India, following a surge in negative

Figure 9.1 Delhi Metro — True Rapid Transport Flies above the Congested Road Traffic.

Source: Peter Newman.

media and a Center for Science and Environment (CSE) study looking at mortality and morbidity in Indian cities suggesting that the Indian government was responsible, ruled that the Delhi government had to create an action plan to control the city's air pollution. The Delhi government presented its first action to the court to combat air pollution in Delhi on December 1996. From this, in July, 1998, the Supreme Court of India introduced a CNG program in Delhi to reduce air pollution. The program stipulated that all buses in the city should run on CNG fuel by 2001 and that taxis and auto rickshaws post-1990 should be replaced by vehicles running on CNG. Furthermore, the program provided financial incentives for taxis and three-wheelers that ran on CNG.

Following the introduction of the program, CNG was introduced in 10,200 buses, 52,623 three-wheelers, 10,350 private cars, 4,497 mini buses, 5,043 taxis, 5,909 light commercial vehicles as well as 689 commercial vehicles. This is a remarkable switch. All public transport in Delhi now runs on CNG and the pollution levels came down as the

Figure 9.2 New Delhi's Metro in the Shadow of a Temple Dedicated to Lord Hanuman, the Hindu Monkey God.

Source: Moore, 2008.

Figure 9.3 CNG Vehicles in Delhi.

Source: Peter Newman.

Figure 9.4 People Coming Together to Combat Pollution.
Source: Delhi Transport Department, 2011.

policy was implemented (Narain and Krupnick, 2007). Unfortunately pollution levels have now risen drastically to levels well above the previous levels due to the rise in private cars and trucks using diesel (Pandey, 2012).

Delhi has been trying to create a sustainable transport city and to provide an example for other cities. The Delhi government is demonstrating further initiatives by introducing a BRT system in Delhi, currently in its first phase. Furthermore, the Delhi Transport Authority proposes an integrated transport network in Delhi which if implemented would enable Delhi to be a far more sustainable transport network.

9.3. Shanghai

Shanghai is one of the largest cities in the world, and the largest in China. It is an important port city located in the estuary of the Yangtze River and has long been the largest economic center in China (Figure 9.5). Shanghai is 6,340.5 km^2, including 6,218.65 km^2 of land and 121.85 km^2 of water. From south to north the city spans approximately 120 km

Figure 9.5 Shanghai Skyline.

Source: Adapted from Fisher, 2012.

and from east to west it spans 100 km. Administratively, Shanghai is divided into 18 districts and 1 county. By the end of 2009, Shanghai's registered urban population reached 13.71 million, while the total number of inhabitants staying in Shanghai for more than six months totalled 19.21 million. Shanghai's development objectives include the creation of international economic, financial, trade and shipping centers, and to be a modern international metropolis.

Shanghai boasts one of the best transportation systems in China with 11 metro lines, 1,100 bus lines, 50,000 taxis and 20 ferry lines. Over the past ten years, Shanghai, following a policy of encouraging public transport in the city center, invested an average of 2.9% of its GDP annually on transport infrastructure. Of this, 41% of the investment was in the metro system. The huge investments in transportation were only possible due to the active role of the Shanghai government and the involvement of the public sector.

The development of the Shanghai transport system across different periods of time illustrates how the city has transformed from a private transport oriented development model to a more sustainable transport oriented model. This has influenced the structure of the whole city and the life of its citizens. The transformations can be divided primarily into a freeway construction period and a metro construction period.

9.3.1. *Freeway Construction Period*

From the 1980s to 1997, Shanghai faced rapid urbanization as the economic boom in China was in full force. To support the rapid economic growth and development, the government started to build road infrastructure. At that time, the Shanghai government was impressed with the sound road network and the 'comfortable life' in American cities where citizens had private cars. The increasing economic wealth in Shanghai meant that many of its inhabitants want to own a private vehicle and the city government through policies and large-scale infrastructure building encouraged higher vehicle ownership. In just ten years the length of roads increased by 40% and the number of cars rose to just over a million. By the end of 2000, the total mileage of city roads in Shanghai had reached 12,227 km, up by 84% over the corresponding 1985 figure. Of this, 7,200 km were expressways and arterial roads.

However, only a small number of citizens benefited from this infrastructure program, as the majority of citizens could not afford a private vehicle and many continued to use public transport and bicycles as their main transport mode. In 1993, the buses had an average speed of 8 km/h and were very crowded. On some of the really congested roads, walking was faster than the bus. Commute times were very long. For people that lived in the Pudong area (east of Huangpu River) but worked in the Puxi area (west of Huangpu River), the commute time could be as long as 3 hours.

9.3.2. *Metro Construction Period*

Public discontent with Shanghai's transportation system grew. The government started to realize that the fast growth of car ownership required large amounts of road space and that no matter how extensive the highway system was, it would not be able to cope with car growth. Construction of new roads in Shanghai lessened the traffic congestion for a short period of time; however the city became increasingly congested. Road congestion was particularly bad during rush hour. In order to support economic and social growth, a healthy and sustainable public transport system was needed. This led the government to recognize that a good public transport system, particularly a metro system, may be the solution.

Large-scale metro construction began in the early 1990s, with the first line opening in 1995. By 2012, there were 11 metro lines in Shanghai

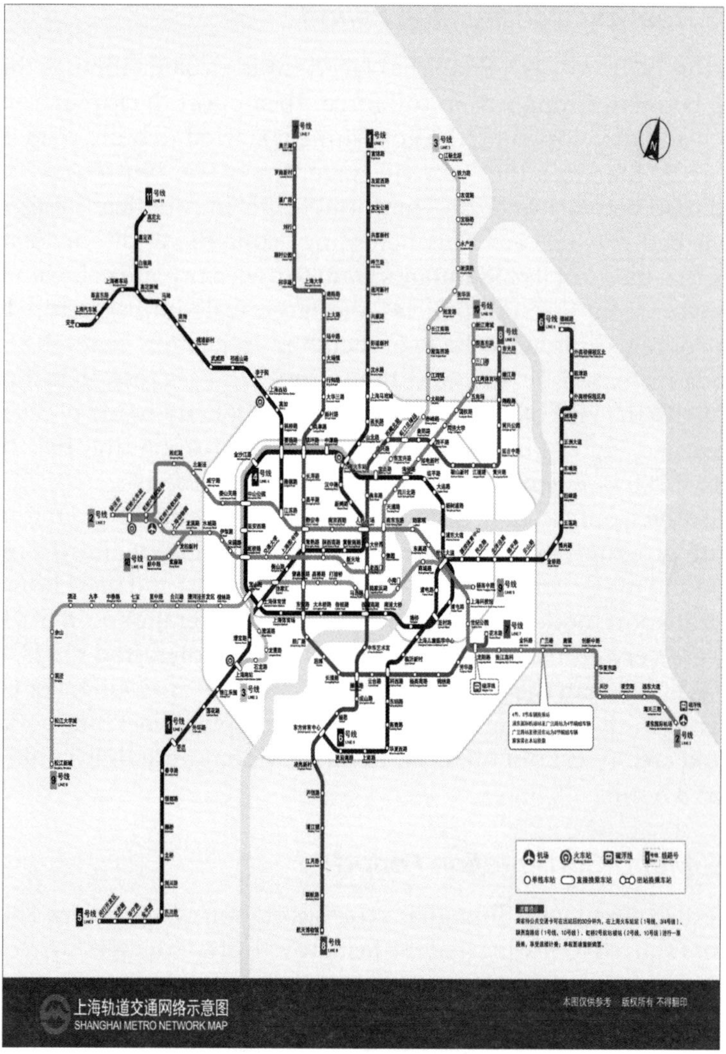

Figure 9.6 Shanghai Metro Network.

Source: Shanghai Shentong Metro Group Co. Ltd, 2010.

with 280 stations and a total operating length of 420 km (Figure 9.6) with a total of 22 lines planned by 2020 The number of passengers serviced by the Shanghai metro system has increased sharply, with an annual increase of nearly 37% between from 2000 to 2005. The total passengers per day rose to around 8 million at the end of 2010. 80% of the developed city area is now covered by the metro and the system is now the largest in the world.

The Shanghai metro system is a remarkable story, transforming the city in a little over a decade. The metro system has reduced the need for cars, and furthermore, has directly and indirectly eased the pressure on the road system. The bus, pedestrian and bicycle networks need to be further integrated with the metro system, however, the metro itself has been a huge achievement, providing the beginning of a more sustainable Shanghai.

9.3.3. *Transport Demand Management Policies*

Despite the introduction of a metro system, Shanghai's motorization has increased rapidly, with the number of motor vehicles increasing from 466,000 to 2,217,000 in nine years, an annual growth rate of 21%. In order to reduce congestion, the Shanghai government has implemented a number of transport demand management policies, including providing priority to public transport, restricting car ownership through license plate auctioning, restricting car parking and motorcycle use. Furthermore, in the late 1990s, the use of unleaded fuel and liquefied petroleum gas (LPG) was promoted in order to reduce air pollution caused by transport. In 2006, this was taken further, with vehicles not meeting emission stands (i.e. those that do not meet European I emission standards) being restricted.

The public transport first policy promotes the continuous increase of the public transport mode share with the aim of making public transport the main mode and private vehicles an 'assistant' mode by ensuring that public transport comes first in land and transportation investments and by ensuring the operational efficiency of public transport. The three-year Action Plan of Developing Public Transport Priority from 2007 to 2009 for Shanghai has been formally put into operation, with an investment of about 110 billion *yuan*. The plan stipulates that all of Shanghai should be accessible with a service ratio of 500 m between public transport stops in the central city and that public transport travel between any two points in the central city will be able to be completed within one hour. The plan also provides more bus services to suburban areas.

The license plate auctioning policy has been in effect since 1994. This policy was implemented in order to control the fast growth of motorized vehicles in Shanghai. In 2006, 68,654 license plates were granted with an average successful bidding price of RMB 46,435. Under the policy, the license is valid for life and can be transferred to other people. The policy has had some unexpected outcomes, such as decreasing the demand for cheaper vehicles and vehicles being licensed in adjacent cities for use in Shanghai. The license plate auction policy has, however, restrained the increase in car ownership. Shanghai has, for example, a lower motorization rate than Beijing.

Shanghai has also regulated vehicle parking charges in order to reduce car use, with charges differentiated in different parts of the city. Within the Central Business District, parking is restricted, with few parking spaces and high parking costs. In other areas more parking spaces are provided at lower costs, particularly in parking lots on the outer city close to transport hubs to encourage car passengers to switch to public transport before they enter the urban center. In the suburban areas, parking is less restricted. Parking facilities are divided into different types: residential parking, workplace parking, public parking, street parking and special bus parking. This allows for consideration of different parking needs. For example, the coordination of workplace parking to the capacity of surrounding roads; the restriction of on-street parking through administrative and economic measures; and the parking needs of buses and around bus depots.

Shanghai has enacted a motorcycle restriction policy. As motorcycles are a very flexible and fast travel mode, their numbers grew very rapidly, reducing road safety amongst other problems. Therefore, the Shanghai government has implemented a policy to gradually reduce motorcycle numbers and has strengthened the oversight of motorcycles. The policy restricts motorcycle use on certain roads within the central city and has increased the traffic regulations regarding motorcycle use. As a result, the negative impact of motorcycles on traffic congestion and order has been reduced.

9.3.4. *Transport during Shanghai Expo, 2010*

The Shanghai Expo in 2010 had a number of green urbanism features in its exhibitions and site design. In particular, the transport plan of the Expo was also aimed at demonstrating green transport innovations whilst ensuring the success of the event and also enabling the whole transport system within the city to benefit. The city used this opportunity to create a multi-modal integrated transport system.

Modeling of the 184 days long Shanghai Expo 2010 predicted around 400,000 visitors to the Expo on an average day, 600,000 visitors on peak days and 800,000 visitors on extremely peak days. Furthermore, visitor distribution by day was predicted to be extremely uneven. Transport demand management was needed to help cope with the transport needs. In order to solve this problem, the transport was carefully planned and managed to ensure minimal car dependence. Five key strategies were used to ensure smooth on-site and off-site transportation: the construction of metro connections, bus-only lanes to and from the city, a water bus and park-and-ride hubs. Rail transport was planned to take care of half the travel to the Expo, with three metro lines serving the area. During the Shanghai Expo 2010, exclusive bus-only lanes were provided for coaches carrying large capacities. Ferry stations were constructed to enable ferry travel to the site, carrying about 5–8% of visitors.

9.3.5. *Conclusion to Sustainable Transport in Shanghai*

Transport is the most fundamental infrastructure in a city. Transport supports the development of a city and the daily life of citizens. The Shanghai government has been working at providing a more sustainable transport system. The benefits are now beginning to be seen in perhaps the most strategic city in the world as China's explosive urban growth now has a model of sustainability and green urbanism that works.

9.4. Bus Rapid Transit and Bike Sharing in Guangzhou, China

Guangzhou, a city in China of 12.7 million people with a population density of 1,708 people per km^2, has done much to implement sustainable transport within the city, including implementing a bus rapid transit (BRT) system and a bike sharing scheme to complement the metro system (Institute for Transportation and Development Policy, 2010; United Nations, Bureau International des Exhibitions and Municipal Government of Shanghai, 2011). The BRT system opened in February 2010, and consists of 980 buses, 26 stations and 23 km of dedicated trunk lines covering 273 km (two-way roads). Travel on the BRT is on one fare and the system in 2011 moved 843,000 passengers per day, with peak passenger flows of 27,400 passengers per hour per direction and with 350 buses moving per hour per direction at peak times. The system has resulted in 30% higher bus speeds with buses traveling at 17–19 km/h during peak times, and an average saving of 6.63 minutes per BRT trip, or 30 million passenger-hours per year. The system has been very popular with bus passenger satisfaction rising from 29% in December 2009, before the BRT was implemented, to 65% in 2010, after the BRT. In addition to the BRT system, a bicycle sharing system was designed to complement the metro system. The bike sharing system opened in June 2010, with 5,000 bicycles and 113 bike stations, mostly located adjacent to the BRT stations. Plans are in place to increase this to 15,000 bicycles, to increase the bike infrastructure and also to implement 5,000 bicycle parking positions at BRT stations.

9.5. Sustainable Transport Momentum in Yogyakarta, Indonesia

Yogyakarta, a major city in central Java, Indonesia, is renowned for being a center of arts, culture and education. The area is home to approximately 2,389,200 people in the metropolitan area (in 2010). Like many cities in the Asian regions, Yogyakarta has been experiencing rapid urbanization, particularly in the form of rural to urban migration. This has resulted in increased traffic congestion and increased air pollution. In order to help combat these issues, the city has been working on various schemes to improve its transport infrastructure. In particular, the city has been looking at implementing a pedestrian-only street in the popular Malioboro area aimed at stimulating revitalization. Furthermore, the city is facilitating cycling and is creating a BRT system.

Malioboro is the urban center of Yogyakarta and home to much of the city's cultural and business activities. Pustral UGM (Center for Transportation and Logistic Studies, Gadjah Mada University), along with the former mayor, proposed a Malioboro Traffic Management and Pedestrianization Scheme for the area in order to revitalize the city center, reduce congestion and localize air pollution and provide pedestrian-only areas. The plan is to redevelop the Tugu Railway Station, pedestrianize Malioboro and also redevelop strategic sites in the area. Modeling of the scheme by Pustral UGM determined that implementation would result in a reduction in delays by 18 minutes per vehicle, reduce congestion severity, reduce productivity loss accounting for 6.9 billion IDR per year and reduce air pollution by as much as 38.5% in the area (Zudianto and Parikesit, n.d.).

Various schemes have been implemented to improve bicycling in the area, including developing a 'Green Map' showing cycling routes, a 'Bikes for School and Work' program and improving cycling at the Gadjah Mada University campus. Green Map has developed a cycling map of the city, which includes highlighting the bicycle-friendly routes that link important destinations throughout the city. The '*Sego Segawe — Sepeda Kanggo Sekolah lan Nyambut Gawe*' (Bikes for School and Work) program was established in 2008 by the Sultan with the objective to increase the mode share of bicycle trips to 10% of total trips through educational and entertainment programs. Gadjah Mada University is the oldest university in Indonesia and has an 'Educopolis Campus' program designed to create a conducive, safe and peaceful campus. This program has included the provision of 'green' bikes, a transport identity card, reorganizing the streets to enable more cyclists to move freely, providing a campus bus and a free parking service. This program is still underway and has had mixed results.

The Department of Transport implemented a Bus Rapid Transit system, '*Trans Jogja*', in the city in 2008. The system has six lines running on the main streets of Yogyakarta, extending from the Jombor bus station in the north to the Giwangan main bus terminal in the south and the Prambanan bus shelter in the east (passing the Adisucipto International Airport). Although the system does not travel on dedicated bus lanes it has been very popular, operating from 6.00 am to 10.00 pm daily.

9.6. Bike Sharing: C-Bike, Taiwan

Public bicycle sharing systems are becoming increasingly popular and can complement public transport systems, enabling systems to have a further reach. Bicycle sharing systems are rapidly being implemented around the world, for example, the Vélib system in Paris and the system in Guangzhou, China. Bike sharing systems work on a simple concept with people borrowing a bike from one station and returning it to another. They are ideal for short trips. They are easy to operate and able to be integrated into existing transport systems. C-Bike is the first bike sharing system in Taiwan.

Taiwan is known for scooters and busy streets. Contradicting this is Kaohsiung City, located in Taiwan's southwestern region. The Taiwan Government has put an extensive effort into promoting biking and Kaohsiung City is no exception. The idea was to promote a 'Green City' for residents and to prepare the city for the 2009 World Games. Kaohsiung City is the first city in Taiwan to offer a bicycle sharing system. The system, called C-Bike, consists of 50 stations, 4,500 bicycles and 180 km of bicycle paths covering the city center, major destinations, metro transport stations and major streets. The system is self-serviced and available 24 hours a day. Like many of the other bike sharing systems, it operates through an electronic membership subscription. The bikes are free for the first 30 minutes with charges occurring after this time, encouraging people to use the bikes for short trips. The system was constructed on a 'build-operate-transfer (BOT) basis' for a cost of NT$90 million ($2.58 million), with NT$15 million ($444,000) contributed by both the Environmental Protection Administration's air pollution control fund and the city government, and NT$60 million ($1.77 million) from the central government's economic stimulus package. The system is operated by Tung Li Development Co.

The C-Bike system has been very popular with the total riding time increasing from 5,433 hours in April 2009 to over 30,000 hours in December 2009. By the end of 2009, it was estimated that the C-Bike system had resulted in the reduction of 200 tons of CO_2 emissions within the city. The system has inspired the YouBike share system in Taipei and the 'Nubija' (Nearby Useful Bike, Interesting Joyful Attraction) system in Changwon, Korea (Figure 9.7).

9.7. The 'Walking Plaza' of Pune, India

> *The beauty of feet moving one before the other, taking me this way and that way, gave me a sense of belonging to this earth that zipping across the city in a vehicle can never provide.*
> *The image of a city is the experience of its streets. If the streets are ugly, then the city is ugly, if the streets are appealing, the city is appealing! If a populace does not fight for its pedestrian domains, then it has lost its culture and its identity!*
>
> Ar. Christopher Benninger, 2010 (urban designer, Pune).

Figure 9.7 The Nubija Bike Share System in Changwon, Korea.
Source: Peter Newman.

Mahatma Gandhi Road (MG Road) in Pune, India, was a very congested commercial road. Globally there is a trend towards reclaiming spaces in the city center that were for cars and car parks for people, as seen in Melbourne, Bogotá, Copenhagen and New York. The economic potency and vibrancy of walkable city centers is now well recognized. Following this trend, MG Road was converted into a vehicle-free walking plaza on Saturdays and Sundays from 4.00 to 10.00 pm. The idea for this conversion emerged from a clean air directive by the Supreme Court and a campaign by the Pune chapter of INTACH[1] (The Indian National Trust for Art and Cultural Heritage) to have certain roads free of vehicles. In 2003, the Central Pollution Control Board declared Pune to be the 16th most polluted city in India. A committee was formed to monitor air pollution. During a review meeting of the committee along with various municipal corporation stakeholders in Pune in October 2005, committee member Sunita Narain urged the Municipal Commissioner to declare MG road (under the control of the Pune Cantonment Board) and Laxmi Road (under the control of the Pune Municipal Corporation)

[1] INTACH is India's largest nationwide non-governmental organization (NGO) committed to the cause of conserving India's rich and diverse natural and man-made heritage.

as pedestrian-only zones, at least during festivals. This idea was taken up and in January 2005, the Pune Cantonment Board (PCB) declared its intention to make MG Road into a 'Walking Plaza'.

Almost a year after it was declared, the project began to be implemented. This included implementing footpaths, street lighting and beautification elements, along with resurfacing of the road. Car parking space around MG Road was augmented with pay-and-park schemes. Management of the Walking Plaza was given to an event management company who ensures that vehicles parked on MG Road were towed away, that barricades on either side of the Walking Plaza were erected and that the roads were kept clean. PCB also allocated ten spots on the road for advertising display boards, to generate revenue for the event management company.

Initially the shopkeepers were against the 'Walking Plaza' idea as their business on Saturdays dropped by 80% and the shopkeepers' association demanded the withdrawal of the road closure. However, PCB was determined to make it a success and surveyed citizens, finding that 80% appreciated the Walking Plaza. Between 10,000 and 20,000 people patronize the Walking Plaza during weekends. Families and friends chat and move around the plaza, children play, trees are decorated with festive lights and the shops are now attracting more people than they had before the closures (Figure 9.8). Some eateries set up chairs and tables on the street. Wi-Fi has been installed in the area. Furthermore, the Walking Plaza has had direct envrionmental benefits. Sound and air pollution tests during the walking hours showed a 40–50% reduction in air pollution and 32–40% decrease in sound levels.[2]

9.8. Janmarg, 'The People's Way': The Ahmedabad Bus Rapid Transport System

BRT systems are becoming increasingly popular particularly when planned in conjunction with other transport modes such as bike sharing schemes. BRT is "a flexible, rubber-tired form of rapid transit that combines stations, vehicles, services, running ways and [intelligent transportation systems] elements into an integrated system with strong identity" (TCRP, 2003). A successful example of a BRT system is 'Janmarg', in the city of Ahmedabad, India (Asian Development Bank,

[2]The MG Road Walking Plaza has been temporarily closed due to security concerns.

Figure 9.8 Pedestrians Favored in Pune Street.

Source: Carolina Tuerel.

2010). This is India's first full BRT system, and was the 2009 winner of the Best Mass Transit Project awarded by the Government of India and the 2010 winner of the Sustainable Transport Award awarded by the Institute for Transportation and Development Policy.

The Ahmedabad government realized that it was only with a long-term vision that the city could be able to sustain its growth whilst enhancing the quality of life of its citizens. Since the most successful cities are those that have a high quality public transport system, the government adopted a sustainable transport system as its primary vision, able to enhance the quality of life of its residents and increase the competitiveness of the city. To fulfil this vision, they choose to implement a BRT system, calling it '*Janmarg*', the people's way.

In 2005, the State Government of Gujarat declared an Urban Development Year and since then has been concentrating on improving the urban environment. Ahmedabad, with a population of more than 5 million people, had been experiencing rapid growth in car ownership and declines in public transport use and walking. Thus, the city was experiencing increased congestion and air pollution. Furthermore, the land use of the city was segregated, with residential uses primarily in the western part of the city and industrial uses primarily in the eastern part.

The project was influenced by the successful BRT systems in Bogotá (TransMilenio) and Curitiba. However, its planning and design was adapted to suit Ahmedabad's specific trends and conditions. A report by the Center for Environmental Planning and Technology (CEPT) to the Gujarat Infrastructure Development Board determined that it was important to consider the high number of low-income people, 55% of which did not have access to private motor vehicles and were dependent on public transport, when planning a BRT system for Ahmedabad. Furthermore, the report maintained that pedestrian and bicycle-friendly streets were vitally important due to the large numbers of pedestrians and cyclists. CEPT maintained that the BRT system could be designed to enhance the compactness of the city, connecting land uses (Centre for Environment Planning and Technology University, 2011; Gujarat Infrastructure Development Board, 2012).

The Janmarg buses run on dedicated lanes along the center of the road with seperated paths for cyclists and wide footpaths for pedestrians. Janmarg has a network of 89 km throughout the city, connecting the central city with the outlying industrial and residential areas as well as institutional areas. The project was developed with an estimated cost of about Rs. 1,000 Crores, and received support under the National Government's Jawaharlal Nehru National Urban Renewal Mission (JnNURM) Scheme. The share of central, state and local government is in the proportion of 35:15:50. The BRT lines were planned based on the land use, density and the areas of maximum accidents.

Initially, the BRT carried around 24,000 passengers for an average stretch of 12.5 km (from the Regional Transport Office to Chandranagar). Later, the system was extended, increasing the number of passengers per day to about 35,000 on weekdays and 40,000 on weekends. The service has a frequency of 4 minutes during the designated peak hours (8.30 am to 11.30 am and 5.00 pm to 8.00 pm) and operates from 6.00 am to 11.00 pm. The average speed of the BRT buses is 26–29 km/h with vehicle utilization of about 207 km per day. The buses are low-floor (0.9 m above ground) allowing for easy access. The stations are well-lit and are equipped with synchronized automatic sliding doors. Digital displays are fed by a centrally-controlled Intelligent Transport System (ITS) linked to the Global Positioning System (GPS) on board each BRT bus. These displays communicate to the waiting passengers when the next few buses are expected to arrive. Pre-board ticketing and fare collection are carried out at ticket counters located at both ends of each station. Janmarg bus stations feature passive solar designs, an inexpensive way to keep the stations naturally cool.

In seven months 34% of commuters in that corridor shifted from private cars to the BRT system. The BRT system reduced carbon emissions by 3,700 tonnes. 400 more CNG buses are planned and officials have a plan to buy carbon credits on that basis as well. The land use patterns around the BRT line are hoped to change although the changes so far show reductions in residential activity and small increases in services and light industries. Many vacant land and slums have been converted to public spaces.

The ex-mayor of Bogotá, Enrique Peñalosa, visited the city and traveled on Janmarg. He appreciated the efforts being made and at the same time suggested some points on which the government could do to improve the system. The BRT system in Ahmedabad shows how a vision based on public transport can help restructure a city.

9.9. E-jeepney: A Green Solution by a Developing City

E-jeepneys are an entirely electric transportation system powered by clean energy. The biggest mass application of the e-jeepneys in Asia is Makati's Green Route (MGR), which operates ten e-jeepneys on the Legazpi and Salcedo routes for free (Institute for Climate and Sustainable Cities, 2009).

E-jeepneys are part of the Climate Friendly Cities Program of the Institute for Climate and Sustainable Cities (ICSC). This program has three major components: a biodigester fed with biodegradable household wastes to generate electricity, a depot or a terminal and a fleet of e-jeepneys that utilize this electricity. The jeepney was originally a product of the United States military, however after World War II, many of the jeepneys were left over. Hundreds of these were sold to the Philippines. Since then, the jeepneys have become known for their flamboyant decoration and crowded seating. They have become a symbol of Philippine culture. However, after six decades of use, the jeeps were still being used, creating environmental hazards. Their huge diesel-powered motors are one of the major contributors to air pollution and ensuing health problems in Philippine cities.

An initiative put forward was to turn these jeepneys into electric jeepneys. Red Constantino, Director of the Institute for Climate and Sustainable Cities, a Manila-based non-governmental organization (NGO), maintained that the old jeepneys were inefficient, producing pollution. So along with a small collection of politicians, business groups and other NGOs, he started a campaign to replace the jeepneys with e-jeepneys. The pilot project was in Makati. This project was crucial in providing a showcase for the future. His intention was to have solutions on the ground to show people that the project was viable. The ultimate goal was to eventually replace all the public utility vehicles with appropriate electric ones.

E-jeepneys were the brainchild of Green Renewable Independent Power Producers, Inc. (GRIPP), in partnership with Mr Robert Puckett, President of Solar Electric Company in the Philippines and supported by Greenpeace. Four e-jeepneys were launched by Makati Mayor Jejomar Binay on 2007. The jeepneys are manufactured locally by the Philippine Utility Vehicles company.

The e-jeepneys are designed to carry 17 passengers and can run for about 120 km on an 8 hour charge at a maximum speed of 35–60 km/h.

The impacts of the e-jeepneys are spreading all over the nation. Puerto Princesa, the capital of Palawan Island, is aiming to introduce an e-jeepney fleet, however, their major ambition is to replace the city's 4,000 gasoline-powered tricycles with electric 'e-trikes'. The momentum has also started to build in Makati, with an intention to open a third green route, Rockwell loop, by May 2010 and other city governments in Metro Manila are placing orders to buy e-jeepneys.

The next big step for Puerto Princesa and Makati is to build biogas plants so that the e-vehicles could be powered by organic waste from local markets and households. In February 2010, Puerto Princesa began construction of a 1 MW biogas plant, costing 2.4 million USD, to fuel its electric public transport fleet.

Since its launch in July 2008, e-jeepneys have been used by schools, resorts, theme parks, industrial zones, Local Government Units (LGUs) and other entities. The project is expected to dramatically reduce the noise and air pollution in Makati.

The e-jeepney project has been a great success and a new green model for transportation, especially for Asian cities (Figure 9.9). This project was initiated by a public-private partnership, which came

Figure 9.9 E-jeepney in Makati.

Source: Institute for Climate and Sustainable Cities, 2011.

together with the aim of making the city greener. They were all convinced that the old diesel jeepneys should be replaced. They had a clear agenda and a long-term vision.

9.10. Conclusion

Transportation choices impact how a city is shaped and how it develops. Public transport, bicycling and walking are all important sustainable modes. Furthermore, as we have seen from this chapter, alternative fuels are important to enable green urbanism to flourish. It is also critical for cities to match their transportation to the needs of the particular city, such as the e-jeepney program in the Philippines, the BRT in Ahmedabad and the pedestrian innovations in Pune.

Many Asian cities are embracing electric metro lines. The Delhi metro carries 1.20 million people per day and has been a lifeline for the Delhi people, enabling access through the city and for businesses to flourish in once hard to access areas. Shanghai has dramatically improved its sustainablity too by implementing a metro line that carries 8 million passengers per day. Transportation demand management policies, such as license plate auctioning and a public transport priority policy have helped enable this increase in ridership. Together, these projects are models for Asia and for the rest of the world.

The hardest part of the transport system to improve is walking and cycling even though it is the cheapest option. That is why the successful demonstrations of Kaohsiung City's popular bike sharing program and the 'Walking Plaza' concept in Pune City, India, are so important.

So many other sustainable transport demonstrations could have been chosen from across Asia but those discussed give a flavor of the green urbanism innovations that are successfully emerging in Asia. The most critical element of Asian cities transport is that the cities are dense and compact. This will enable them to be resilient in the Age of Scarcity when oil prices will make highly car-dependent, low-density cities so much more vulnerable. They will need to continue to make innovations in sustainable transport as will all cities, but Asian cities are much better structured to be able to respond to this agenda.

CHAPTER TEN

Conclusions: Will Asian Cities Take Over the Green Urbanism Agenda?

> *The Asian and Pacific region is also leading the globe in commitments to green investments, including low-carbon power generation (renewable energy and carbon capture and storage), energy and fuel efficiency (buildings, public transport and electricity grids), and water supply and waste management* (United Nations and Asian Development Bank, 2012, p. xiii).

This book has examined green urbanism in Asia from seven perspectives:

1. ***Renewable Energy Cities***:
   Dramatic increases in the use of photovoltaics (PV) in Chinese, Taiwanese and Japanese cities have been seen along with the first fully PV-powered city of Masdar. Wind has become the renewable energy source of choice in many Indian, Chinese, Korean and Philippine cities, whilst geothermal is being demonstrated in Philippine and Chinese cities. Traditional biomass sources are also being made part of the Asian renewable city. China, India and other parts of Asia are now investing more in renewable energy than the whole of the developed world.
2. ***Bioregional Carbon Neutral Cities***:
   Across Asia there are thousands of bioregional examples emerging of how cities are recognizing the need to tap local bioregional resources not just to enhance biodiversity and recreation but to link into carbon offset programs. Although this is small compared to the growth of

carbon neutral cities in the developed world, it is anticipated that these initiatives will rapidly grow in Asia.

3. ***Distributed Cities***:
   Asian cities are ideally placed to develop small-scale infrastructure solutions for energy, water and waste. Tianjin Eco-city in China is developing global leadership in its smart grid to manage local power and water. Marina Bay in Singapore is a small but iconic example of how to do it but many examples are also emerging in places like Jakarta and Dhaka that are low-cost decentralized systems. All the low-carbon cities of China have distributed infrastructure. Considerable growth in this approach to city building is anticipated in Asian cities.
4. ***Biophilic Cities***:
   The need for high-density cities in Asia now requires high density greening solutions such as biophilic urbanism. This set of innovations lead to major increases in urban biodiversity, cooling urban atmospheres and other green benefits, and is now being demonstrated in many cities across Asia, particularly in new urban developments in Singapore. The Khoo Teck Puat Hospital, 158 Cecil Street, Hougang Primary School and Bishan Park are rapidly achieving iconic status on a global stage, along with the dramatic supertrees of the Gardens by the Bay at Marina Bay.
5. ***Eco-efficient Cities***:
   Asian cities are very good at this, probably because of their density and rapid industrialization. The case studies show how whole cities like Singapore and cities in Japan and South Korea, as well as parts of cities can be demonstrations of 'closing the loop' on urban metabolism. The use of recycled building materials has been demonstrated in Asian urban slums as well as in iconic projects like Indonesia's 'Trashion' products which show beauty rising from the garbage. Chinese industrial ecology and eco-efficiency is now being demonstrated as the world's best practice.
6. ***Place-based Cities***:
   Asian culture is deep and geographically distinct but has also produced in recent decades some of the worst examples of modernist mass-produced high-rise buildings to satisfy urban housing growth pressures. This is certainly not something Asia has to itself. Reclaiming 'place' through heritage projects, public art, natural features and local architectural styles are now happening in many places,

especially through water-sensitive urban design such as the iconic river restoration in Seoul.

7. ***Sustainable Transport Cities***:
   Dense Asian cities are already much more sustainable than many other cities in their transport due to short distances that enable walking, cycling and public transport. However, the book has shown that innovations in metro lines, bus rapid transport (BRT), ferries, e-jeepneys, 'Walking Plazas' and many more, are also appearing in Asian cities.

The question remains: Will these Asian green urbanism innovations become so mainstreamed that they will show the world how to build a new green economy in the 'Age of Scarcity'?

The answer may well be yes for the following reasons:

1. The growth pressures in Asia are clearly now dominating global urban demographics. The previous places which drove the waves of innovation were places where population pressures created the need for new efficiencies, new ways of creating urban housing and infrastructure and new ways of doing business. As the UN and the Asian Development Bank state "[a]bout two-thirds of the $8.0 trillion needed for infrastructure investment in Asia and the Pacific between 2010 and 2020 will be in the form of new infrastructure, which creates tremendous opportunities to design, finance and manage infrastructure according to principles of sustainability, accessibility, eco-efficiency and social inclusiveness" (United Nations and Asian Development Bank, 2012, p. 54).
2. The rapid processes of change in Asian cities indicate that change to green urbanism can be achieved quickly. The demonstrations in cities like Shanghai which built the world's biggest metro in a decade, or Singapore which has dramatically increased it biophilic qualities in a few years, stand out on a global stage. These changes are driven not just by demography but also by a desire to lead the world. The Chinese industrial ecology demonstrations are another example of this rapid implementation of ideas that are seen elsewhere as worthwhile but are generally not done.
3. More than anything else sustainable transport will be a key to how soon global cities can emerge with new economies in an age of oil scarcity. The structure of Asian cities will enable this rapid transition.

They are not built for cars so when rapid transit like metros and BRTs are built or when pedestrian and cycle facilities are provided, the city can dramatically reduce its oil dependence. The signs of this happening are everywhere now.

4. Culturally, the cities of Asia are highly community-oriented. The transference of ideas about sustainability into this deep Asian world view is likely to lead to a strong political drive for green urbanism. Evidence for this emerging in Asia where the Nielsen Global Survey of Corporate Citizenship in 2011 covering 28,000 consumers across 56 countries found that Asia–Pacific came out the top region for consumers willing to buy brands that made an attempt to "give back to society" (Nielsen, 2012). On top of this, the number one cause people wanted to support was environmental sustainability.

Whether Asian green urbanism will lead the world is likely to depend on many factors like leadership qualities by urban politicians, the activity of non-governmental organizations that enable media campaigns to highlight projects, active involvement of universities, business leaders prepared to take a risk on green urbanism, and the day-to-day choices of ordinary householders.

We have been encouraged to believe that green urbanism in Asia is rapidly emerging as a global leadership force.

References

Adivasi Academy (2006) *Adivasi Academy*, http://www.adivasiacademy.org.in

Aga Khan Award for Architecture (2010) Tulou Collective Housing, *Aga Khan Award for Architecture*, http://www.akdn.org/architecture/project.asp?id=3860.

Alusi, A, RG Eccles, AC Edmondson and T Zuzu (2011) Sustainable Cities: Oxymoron or the Shape of the Future? *Working Paper, Harvard Business School*, http://www.hbs.edu/research/pdf/11-062.pdf

Amir, HH (2000) Community Participation in the Development of Permaculture in Sukabumi, Indonesia, Index of Case Studies, *Sustainability and Technology Policy, Murdoch University*, http://www.istp.murdoch.edu.au/ISTP/casestudies/Case_Studies_Asia/sukabumi/sukabumi.html

ArchNet (n.d.) *ArchNet: Islamic Architecture Community*, Aga Khan Trust for Culture, http://archnet.org

ARCPROSPECT International Foundation (2011) Tierra Design. Changi Airport Terminal 3: The Green Wall, *Arcprospect*, http://www.arcprospect.org/index.php?option=com_content&view=article&id=2210%3Atierra-design-changi-airport-terminal-3&catid=87%3Alandscaping&Itemid=14&lang=en.

Arup (2009) Beijing Changxindian Low Carbon Community Concept Plan — Pioneering the Development of Low Carbon Zoning Codes in China, *2009 Isocarp Award for Excellence Project Brief, Arup*.

Arup (2011) Wanzhuang Eco-city, *Arup*, http://www.arup.com/Projects/Wanzhuang_Eco-city.aspx

Asia 3R Conference (2006) Integrated Solid Waste Management in Singapore, *National Environment Agency and Ministry of the Environment and Water Resources, Singapore*.

Asia Biomass Office (2008) Thailand — The Biggest Biomass Fuel Producer of Southeast Asia, *Asia Biomass Office, Asia Biomass Energy Cooperation Promotion Office*, http://www.asiabiomass.jp/english/topics/1006_03.html

Asian Development Bank (2010) Urban Development Policy of India (Part 1) Mobility of People by Bus Rapid Transit, *Urban Innovations and Best Practices*, ADB, http://www.adb.org

Asian Development Bank (2012) *Key Indicators for Asia and the Pacific 2012, Special Chapter: Green Urbanization*, http://www.adb.org/sites/default/files/pub/2012/ki2012-special-chapter.pdf

Asian Development Bank (n.d.) *Bus Station Locations* [presentation], http://www.adb.org/Documents/Urban-Development/bus-station-location.pps

Auroville (2010) *Auroville*, http://www.auroville.org

Averositi Holding (2009) *Zira Island*, http://www.ziraisland.com/

Banerjee, R (2009, 13 September) Thyagaraj Stadium to Generate Own Power, *The Times of India*, http://timesofindia.indiatimes.com/city/delhi/Thyagaraj-Stadium-to-generate-own-power/articleshow/5003989.cms

Banga, N (2010, 4 April) Thyagaraj Stadium: India's First Eco-friendly Sports Complex, *Merinews*, http://www.merinews.com/article/thyagaraj-stadium-indias-first-eco-friendly-sports-complex/15802998.shtml

Bareja, B (2010) Intensify Urban Farming in the City by Growing Crops, *Crops Review*, http://www.cropsreview.com/urban-farming.html

BCA GreenMark (n.d.) Singapore's First Zero Energy Building at the BCA Academy, *BCA GreenMark*, http://www.greenmark.sg/about_proj_zero.html

Beatley, T (2000) *Green Urbanism: Learning from European Cities*, Washington DC: Island Press.

Beatley, T (2004) *Native to Nowhere. Sustaining Home and Community in a Global Age*, Washington DC: Island Press.

Beatley, T (2005) *Native to Nowhere. Sustaining Home and Community in a Global Age*, Washington DC: Island Press.

Beatley, T (2011) *Biophilic Cities: Integrating Nature into Urban Design and Planning*, Washington DC: Island Press.

Beatley, T and K Manning (1997) *The Ecology of Place*, Washington, DC: Island Press.

Beatley, T and P Newman (2009) *Green Urbanism Down Under: Learning from Sustainable Communities in Australia*, Washington DC: Island Press.

Beattie, C, J Bunning, J Stewart, P Newman and M Anda (2012) Measuring Carbon for Urban Development Planning, *The International Journal of Climate Change, Impacts and Responses* 3(4): 35–52.

Benedict, M and E McMahon (2006) *Green Infrastructure: Linking Landscapes and Communities*, Washington, DC: Island Press.

Benninger, C (2010, 5 September) The Beauty of Walking, *Pune Mirror*, http://www.punemirror.in/index.aspx?Page=article&sectname=Columnists%20-%20 Christopher%20Benninger&sectid=124&contentid=2010090520100905014223981375cfe8b

Biggs, A (2010, 28 June) Seoul Gets Image of Soft City with Cheonggyecheon, *The Korea Times*, http://www.koreatimes.co.kr/www/news/biz/2010/07/291_68399.html

Bloomberg New Energy Finance (2012) *Bloomberg New Energy Finance*, http://go.bloomberg.com/

Bureau of Energy (n.d.) Sustainable Development of Renewable Energy, *Bureau of Energy, Ministry of Economic Affairs*, http://www.moeaboe.gov.tw/

Burwell, D (2005) Way to Go! Three Simple Rules to Make Transportation a Positive Force in the Public Realm, Making Places, http://www.pps.org

C40 Cities (2011) *C40 Cities*, http://live.c40cities.org/

Campbell, M C and D A Salus (2003) Community and Conservation Land Trusts As Unlikely Partners? The Case of Troy Gardens, Madison, Wisconsin, *Land Use Policy* 20: 169–180.

Centre for Environment Planning and Technology University (2011) Corridor Identification — Bus Rapid Transit System — Gujarat Infrastructure Development Board, *Ahmedabad: Centre for Environment Planning and Technology University*, http://www.gidb.org/

Centre National de la Recherche Scientifique and United Nations Convention on Biological Diversity (2010) Rebirth of a River in Seoul (Republic of Korea), Greening the City, http://www.cnrs.fr/cw/dossiers/dosbioville E/biocity.html#

Cervero, R (2008) *Effects of TOD on Housing, Parking and Travel, Transit Cooperative Research Program Report* 128, Washington, DC: Federal Transit Administration.

Chadha, M (2010, 12 May) Indian State to Obligate Utilities to Buy Electricity from Renewable Energy Power Plants, *Eco Politology*, http://ecopolitology.org/2010/05/12/indian-state-to-obligate-utilities-to-buy-electricity-from-renewable-energy-power-plants/

Chiang, K and A Tan (Eds) (2009) *Vertical Greenery for the Tropics*, Singapore: National Parks Board, Singapore Botanical Gardens and Building and Construction Authority.

China Wind Power Center (2012) Installed Wind Power Capacity, *China Wind Power Center*, http://www.cwpc.cn/cwpc/en/node/6425

City of Fremantle (2012) *City of Fremantle*, http://www.fremantle.wa.gov.au

City of Hannover (1998) Hannover Kronsberg: Model for a Sustainable New Urban Development, City of Hannover.

City of Malmö (2005) Sustainable City of Tomorrow: Experiences of a Swedish Housing Exposition, *Stockholm: Swedish Research Council for Environment, Agricultural Sciences and Spatial Planning*.

City of Sydney (2012) Development in 2030: Energy Master Plan, *City of Sydney*, http://www.sydney2030.com.au/development-in-2030/city-wide-projects/powering-sydney-allan-jones

Clean Energy Awards (2007) Winner of the Policy and Lawmaking Category 2007, *Clean Energy Awards*, http://www.cleanenergyawards.com/top-navigation/nominees-projects/nominee-detail/project/64/?cHash=0dc2996ce1/

Clough, L (Topic Ed) (2008) Energy Profile of Philippines. In CJ Cleveland (Ed) *Encyclopaedia of Earth*, http://www.eoearth.org/article/Energy_profile_of_Philippines

Cochrane, J (2010, 23 June) Green Jakarta — Green Water, or Any Color as Long as It Is Clean, *Jakarta Globe*.

Common Room Network Foundation (2011) *Common Room Network*, http:// commonroom.info/about/

Council on Tall Buildings and Urban Habitat (2008) Tall and Green: Typology for a Sustainable Urban Future, *CTBUH 8th World Congress*, Dubai, 3–5 March, 2008, http://www.ctbuh.org/Publications/TechnicalGuides/CTBUH8thConference TallGreen/tabid/645/language/en-GB/Default.aspx

Daddio, D (2009, 7 November) Towards a More Sustainable Jakarta, *The City Fix*, http://thecityfix.com/towards-a-more-sustainable-jakarta/

Delhi Metro Corporation (2010) *Delhi Metro Rail Corporation*, http://www.delhimetrorail.com/

Delhi Transport Department (2011) Delhi Transport Department, *Government of NCT of Delhi*, http://transport.delhigovt.nic.in/

Department of Environment and Natural Resources (2011) Pasig River Rehabilitation Program (Project Component — River Rehabilitation Secretariat) Phase II, *Department of Environment and Natural Resources, Republic of the Philippines*, http://faspo.denr.gov.ph/Prog_Proj/COMPLETED/Pasig%20River%20Rehabilitation%20Program%20II.pdf

Design Scene (2009) Masdar City Centre by LAVA, *Design Scene*, http://www.designscene.net/2009/09/masdar-city-centre-by-lava.html

Detik News (2012) *Detik News*, http://www.detiknews.com

Diani, H (2009, 25/26 July) Water Worries. The Sewage: Poor Sanitation Means Illness and High Costs, *Jakarta Globe*, 25/26 July, pp. 10–11.

Djumena, E (2010a, 20 June) Sampah Plastik Jadi Tas Sayang Lingkungan, *Kompas*, http://bisniskeuangan.kompas.com/read/2010/06/20/09314220/Sampah.Plastik.Jadi.Tas.Sayang.Lingkungan

Djumena, E (2010b, 3 October) Vas Bunga Dari Sampah Plastik, *Kompas*, http://bisniskeuangan.kompas.com/read/2010/10/03/07521629/Vas.Bunga.dari.Sampah.Plastik

Droege, P (2006) *The Renewable City*, Chichester: Wiley.

Druk White Lotus School (2012) *Druk White Lotus School*, http://www.dwls.org/

Duquennois, A and P Newman (2009) Linking the Green and Brown Agendas: A Case Study on Cairo, Egypt, Case Study for UN Global Review of Human Settlements, http://www.worldurbancampaign.org/downloads/docs/GRHS2009CaseStudyChapter06Cairo.pdf/

Entrepreneurship (2010) Tas Cantik dari Sampah Plastik, *Entrepreneurship*, http://www.ciputraentrepreneurship.com/film/1029.html

Ewing, RH, K Bartholomew, S Winkelman, J Walters and D Chen (2007) *Growing Cooler: The Evidence on Urban Development and Climate Change*, Washington, DC: Urban Land Institute.

Farrelly, E (2005, 26 April) Attack of Common Sense Hits Planners, *Sydney Morning Herald*.

Fisher, JP (2012) Tagesansicht von Pudong vom Bund aus. http://commons.wikimedia.org/wiki/File:2012_Pudong.jpg

Fractal Enlightenment (2008, 1 October) Religious Places Go Green, *Fractal Enlightenment*, http://fractalenlightenment.com/793/sustainability/religious-places-go-green

Frumkin, H (2008) Nature Contact and Human Health: Building the Evidence Base. In S Kellert, J Heerwagen and M Mador (Eds), *Biophilic Design* (pp. 107–118), Hoboken: John Wiley and Sons.

Fujita, T (2006) *Eco-town Projects, Environmental Industries in Progress: Environment-conscious Type of Town Building*, Japan: The Ministry of Economy, Trade and Industry.

Future Policy (2007) China's Solar Capital — Success through Government Leadership, *Future Policy*, http://www.futurepolicy.org/2884.html
Galarpe, K (2010, 6 January) 'River Warriors' Fight to Keep Pasig Clean, *ABS–CBN News*, http://www.abs-cbnnews.com/lifestyle/06/01/10/river-warriors-Figureht-keep-pasig-clean
Gardens by the Bay (2011) *Gardens by the Bay*, http://www.gardensbythebay.org.sg/
Garis Hijau (2008) *Garis Hijau*, http://www.garishijau.itrademarket.com
Gehl, J and L Gemzøe (2000) *New City Spaces*, Copenhagen: Danish Architectural Press.
Gehl, J, L Gemzøe, S Kirknæs and BS Søndergaard (2006) *New City Life*, Copenhagen: Danish Architectural Press.
Geng, Y and B Doberstein (2008) Developing the Circular Economy in China: Challenges and Opportunities for Achieving 'Leapfrog Development', *International Journal of Sustainable Development and World Ecology* 15: 231–239.
Geng, Y and H Zhao (2009) Industrial Park Management in the Chinese Environment, *Journal of Cleaner Production* 17: 1289–1294.
Geng, Y, P Zhang, S Ulgiati and J Sarkis (2010) Energy analysis of an Industrial Park: The Case of Dalian, China, *Science of the Total Environment* 408: 5273–5283.
Girardet, H (2000) *The Gaia Atlas of Cities*, London: Gaia Books.
Gizmotrix (2009, 22 May) Taiwan's New Solar Powered Stadium, *Gizmotrix*, http://www.gizmotrix.com/article/150/Taiwans-New-Solar-Powered-Stadium/
Global Environment Centre Foundation (2005) *Eco-towns in Japan: Implication and Lessons for Developing Countries and Cities*, Global Environment Centre Foundation.
GMA News (2010, 15 February) Palawan Biodigester to Produce Clean Energy for Electric Jeeps, *GMA News*, http://www.gmanews.tv/story/183956/palawan-biodigester-to-produce-clean-energy-for-electric-jeeps
Gordon, R (2005, 8 Sept) Boulevard of Dreams, *SFGate*, http://www.sfgate.com
Government of India, Ministry of New and Renewable Energy (2002–2003) *Annual Report 2002–2003*, Chapter 4, http://www.mnre.gov.in/annualreport/2002_2003_English/ch4_pg4.htm
Grameen Bank (1984) *Grameen Bank Housing Programme*, https://archnet.org/library/sites/one-site.jsp?site_id=950
Grameen Bank (2011) *Grameen Bank*, http://www.grameen-info.org/
Green Indonesia (2011) *Green Indonesia*, http://greenindonesia.blogspot.com/
Green School (2012) *Green School*, http://www.greenschool.org/
Gujarat Infrastructure Development Board (2012) *Gujarat Infrastructure Development Board*, http://www.gidb.org/
Haider, MZ (2007) Competitiveness of the Bangladesh Ready-made Garment Industry in Major International Markets, *Asia-Pacific Trade and Investment Review* 3(1): 3–27, http://www.unescap.org/tid/publication/aptir2456_haider.pdf
Halweil, B and D Nierenberg (2007) Farming the Cities. In L Starke (Ed) *State of the World, 2007*, World Watch Institute, Washington, DC.

Han, S (2012, 2 May) South Korean Parliament Approves Carbon Trading System, *Bloomberg*, http://www.bloomberg.com/news/2012-05-02/south-korean-parliament-approves-carbon-trading-system.html

Harden, B (2008, 4 October) Filipinos Draw Power From Buried Heat, *The Washington Post*, Washington Post Foreign Service, http://www.washingtonpost.com/wp-dyn/content/article/2008/10/03/AR2008100303843.html

Hardoy, J, D Mitlin and D Satterthwaite (2001) *Environmental Problems in an Urbanising World*, London: Earthscan.

Hargrove, C and M Smith (2006) *The Natural Advantage of Nations*, London: Earthscan.

Hargroves, K and Smith, M (Eds) (2005) The Natural Advantage of Nations: Business Opportunities, Innovation and Governance in the 21st Century, *Earthscan*, London, Chapter 1, http://www.naturaledgeproject.net/NAON Chapter1.4.aspx

Hawkens, P, A Lovins and H Lovins (1999) *Natural Capitalism: The Next Industrial Revolution*, London: Earthscan.

Her, K (2008, 1 January) An Industrial City Goes Green, *Taiwan Review*, http://taiwanreview.nat.gov.tw/ct.asp?xItem=25050&CtNode=1672

Heringer, A (n.d.) METI — Handmade School in Rudrapur, *Anna Heringer*, http://www.anna-heringer.com/index.php?id=31

Heringer, A and E Roswag (2010, 4 March) Handmade School/Bangladesh, *RADDblog*, http://raddblog.wordpress.com/2010/03/04/handmade-school-bangladesh-by-anna-heringer-eike-roswag/

Higgins, A (2010, 17 May) With Solar Valley project, China Embarks on Bold Green Technology Mission, *The Washington Post*, http://www.washingtonpost.com/wp-dyn/content/article/2010/05/16/AR2010051603482.html?wprss=rss_world&sid=ST201005170109

Himin Solar (2012) *Himin Solar*, http://www.himinsun.com/

Ho, G (Ed) (2002) *International Sourcebook on Environmentally Sound Technologies for Wastewater and Stormwater Management*, United Nations Environment Programme (UNEP) International Environmental Technology Centre (IETC), London: International Water Association Publishing.

Huang, M (2009) China Solar Valley — Renewable Energy Policy in China, Paper presented at *Local Renewables Conference*, Freiburg, Germany 27–29 April 2009, http://www.local-renewables-conference.org/freiburg2009/home/

Hudson, G (2008a, 14 August) South Korean Islanders Ask 'Who Owns the Wind?' *Eco Localizer*, http://ecolocalizer.com/2008/08/14/south-korean-islanders-ask-who-owns-the-wind/

Hudson, G (2008b, 9 May) South Korean Solar System Community on Jeju Island a Brilliant Idea, *Eco Localizer*, http://ecolocalizer.com/2008/05/09/south-korean-solar-system-community-on-jeju-island-a-brilliant-idea/

Index Mundi (2011) Indonesia Population, *Index Mundi*, http://www.indexmundi.com/indonesia/population.html

Indian Power Sector (2011) Wind Power, *Indian Power Sector*, http://indianpowersector.com/renewable-energy/wind-power/

Industrial Technology Research Institute (2011) *Industrial Technology Research Institute*, http://www.itri.org.tw/eng/

Institute for Climate and Sustainable Cities (2011) *E-Jeepney*, http://www.ejeepney.org/

Institute for Climate and Sustainable Cities (ICSC) (2009, 10 November) Electric Jeepney Institute Launches Fare-Free, Emission-Free Makati Green Route, *ICSC*, http://www.ejeepney.org/content/electric-jeepney-institute-launches-fare-free-emission-free-makati-green-route

Institute for Transportation and Development Policy (2010) *Institute for Transportation and Development Policy*, http://www.itdp.org/

International Solar Cities Initiative (2010) Solar Congress Venue, *International Solar Cities Initiative*, Office of the Organization Committee of the 4th International Solar Cities Initiative World Congress 2010, http://www.chinasolarcity.cn/Html/Solar/105609601.html

Jacobs, J (1984) *Cities and the Wealth of Nations*, Harmondsworth, UK: Penguin.

Jain, AK (2008) *The Delhi Metro*, Delhi: Delhi Development Authority.

Jain, S (2004, 26 May) Smog City to Clean City: How Did Delhi Do It? *Mumbai Newsline*.

Jia Han (2009) *Jia Han*, http://bernhan.files.wordpress.com/2009/01/pulau-semakau-landfill.jpg

Justa, A (2009, 10 December) Spectacular Solar Trees: Delight to Your Eyes and the Environment, *Green Diary*, http://www.greendiary.com/entry/spectacular-solar-trees-delight-to-your-eyes-and-the-environment/

Kamal, N and KM Zunaid (n.d.) *Education and Women's Empowerment in Bangladesh*, http://centers.iub.edu.bd/chpdnew/chpd/download/publications/WP-Education%20&%20Women% 20Empowerment-Working%20Paper-N%20 Kamal.pdf

Kaohsiung City Government (2011, 17 January) *Kaohsiung's Transformation into Taiwan's Leading Bicycle Friendly City*, http://www.kcg.gov.tw/EN/NewsArt_Detail.aspx?n=D025BFDE98126962&ss=2FBB442B7B69CD95&parent=DD2A3343D78E64EB

Karlenzig, W and D Zhu (2010) China's Provincial and City Low-carbon Pilot Programs: A New Opportunity for Global Emissions Reductions in Low Carbon Accounting, Management and Credit System, *4th China World Forum for the Shanghai Academy of Social Sciences*, Green Economy, Shanghai, 7 November, 2010, http://www.commoncurrent.com/pubs/Shanghai.WK.11.7.10.pdf

Kellert, SR, J Heerwagen and M Mador (Eds) (2011) *Biophilic Design: The Theory, Science and Practice of Bringing Buildings to Life*, Hoboken, New Jersey: John Wiley & Sons.

Kelley, M (2008, 9 December) The Cheonggyecheon Stream, *DiscoveringKorea*, http://discoveringkorea.com/2008/12/09/cheonggyecheon/

Kelola Sampah (2010, 9 July) Inovasi Tas Daur Ulang Menembus Ekspor, *Kelola Sampah*, http://kelolasampah.blogspot.com/2010/07/inovasi-tas-daur-ulang- menembus-ekspor.html

Kenworthy, J (2008) An International Review of the Significance of Rail in Developing More Sustainable Urban Transport Systems in Higher Income Cities, *World Transport Policy and Practice* 14(2): 21–37.

Kenworthy, J and F Laube (2001) *The Millennium Cities Database for Sustainable Transport*, Brussels: Union Internationale des Transports Publics.

Kenworthy, J, F Laube, P Newman, P Barter, T Raad, C Poboon and B Guia (1999) *An International Sourcebook of Automobile Dependence in Cities, 1960–1990*, Boulder, Colorado: University Press of Colorado.

Koh-Lim, WJ (n.d.) 'Marina Bay towards a High Density Sustainable Waterfront City', URA presentation.

Kompas (2010a) Hidup Sejahtera bersama Sampah, *Kompas*, http://bisniskeuangan.kompas.com/read/2010/11/07/07465493/Hidup.Sejahtera.bersama.Sampah

Kompas (2010b) Trashion, Limbah Plastik Bernilai Ekonomi Tinggi, *Kompas*, http://otomotif.kompas.com/read/2010/02/25/12091349/Trashion.Limbah.Plastik.Bernilai.Ekonomi.Tinggi

Kon Pedersen Fox Associates (2003) *New Songdo City: Green City*, KPF Sustainable Design, New York: http://www.songdo.com/Uploads/FileManager/Songdo/Sustainability%20PDF/KPF%20Sustainable%20Designs.pdf

Korten, D (1999) *The Post Corporate World*, West Hartford: Kumarian Press.

Kostoff, S (1991) *The City Shaped*, London: Thames and Hudson.

Kriscenski, A (2008, 14 January) Solar Ark: World's Most Stunning Solar Building, *Inhabitat*, http://inhabitat.com/solar-ark-worlds-most-stunning-solar-building/

LAVA (n.d.) *Laboratory for Visionary Architecture*, http://www.l-a-v-a.net/projects/masdar-city-centre/

La Mesa Ecopark (2012) *La Mesa Ecopark*, http://www.lamesaecopark.com/

Landscape Forum Nature and Living (2011) A Challenging Green Initiative in the CBD, *Landscape Forum Nature and Living* 6: 13–21.

League of Provinces of the Philippines (n.d.) http://www.lpp.gov.ph

Lee, HL (2010a, 14 May) CLP India Opens 99 MW Theni Wind Farm in Tamil Nadu State, *Recharge*, http://www.rechargenews.com/energy/wind/article214801.ece?print=true

Lee, HL (2010b, 28 June) Geothermal Firm EDC Pursues $150 m Loan, Eyes Wind Project, Geothermal, *Recharge*, http://www.rechargenews.com/energy/geothermal/article218939.ece

Lee, I (2006) Cheonggyecheon Restoration Project: A Revolution in Seoul, ICLEI, Seoul Metropolitan Government, http://worldcongress2006.iclei.org

Lerch, D (2007) *Post Carbon Cities: Planning for Energy and Climate Uncertainty*, Portland, OR: Post Carbon.

Levesque, T (2007, 5 June) Rizhao: China's Solar-Powered Sunshine City, *Inhabitat*, http://inhabitat.com/rizhao-the-sunshine-city/

Lippo Village (2012) *Lippo Village*, http://www.lippovillage.com

Living into the Wild (2012) BR Hills: The Land of Tribal, *Living into the Wild*, http://www.livingintothewild.com/wild-diaries/br-hills-the-land-of-tribal.html

Local Governments for Sustainability (ICLEA) (2004) *SEA–CCP Case Outline No. 5*, Puerto Princesa: Local Governments for Sustainability.

Mah, BT (2006, 5 October) Speech at the Opening Address by Mr Mah Bow Tan Minister for National Development, at the Unveiling of the Marina Bay

Financial Centre, Roof Terrace, Esplanade — Theatres on the Bay, http://www.ura.gov.sg/pr/text/pr06-74a.html
Martinot, E and J Li (2010) Renewable Energy Policy Update for China, *Renewable Energy World*, http://www.renewableenergyworld.com/rea/news/article/2010/07/renewable-energy-policy-update-for-china
Matan, A and P Newman (2012) Jan Gehl and New Visions for Walkable Australian Cities, *World Transport Policy and Practice* 17(4): 30–37.
Matan, A, R Trubka, P Newman and S Vardoulakis (2012) Review of Public Health and Productivity Benefits from Different Urban Transport and Related Land Use Options in Australia, Paper presented at *The 5th Healthy Cities: Working Together to Achieve Liveable Cities Conference*, Geelong, 6–8 June 2012 [Peer reviewed] http://healthycities.com.au/BOP/PR.pdf
McDonaugh, W and D Braungart (2002) *Cradle to Cradle: Remaking the Way We Make Things*, New York: North Point Press.
METI School (2011) *Meti School in Bangladesh*, http://www.meti-school.de/daten/projektbeschreibung_e.htm
Microfinance Information Exchange (2012) *Microfinance Information Exchange*, http://www.themix.org
Ministry of the Environment and Water Resources (2002) *The Singapore Green Plan 2012 (SGP2012)*, Singapore: Singapore Government.
Ministry of the Environment and Water Resources (2007) *Key Environmental Statistics, 2007*, Singapore: Ministry of the Environment and Water Resources.
Ministry of the Environment and Water Resources (2011) *Active, Beautiful, Clean Waters (ABC Waters) Programme Factsheet*, Ministry of the Environment and Water Resources, Singapore: Government of Singapore, http://app.mewr.gov.sg/data/ImgCont/1386/9.%20Factsheet-%20ABC%20Waters%20[web].pdf
Ministry of National Development (2012, 9 January) Newsroom, *Ministry of National Development*, http://app.mnd.gov.sg/Newsroom/NewsPage.aspx?ID=3259&category=Parliamentary%20Q%20&%20A&year=2012&RA1=&RA2=&RA3=
Moore, M (2008, 19 April) China: The New Rulers of the World, *The Telegraph*, http://www.telegraph.co.uk/culture/books/non_fictionreviews/3672728/China-the-new-rulers-of-the-world.html
Mouritz, M (1987) *Sustainable Urban Water Systems* (PhD thesis), ISTP, Murdoch University, Murdoch.
Multiply (2009, 26 February) Luis Liwang, About Kapit Bisig sa Ilog Pasig, *Multiply*, http://ilogmoipatrolmo.multiply.com/journal/item/21
Murphy, C (1999) *Cultivating Havana: Urban Agriculture and Food Security in the Years of Crisis*, Institute for Food and Development Policy, Report No 12. Oakland, CA: Food First Publication.
Murray, S (2010, 7 September) New Tulou: Update of Chinese Dwelling Proves Popular, *Financial Times*, http://www.ft.com/cms/s/0/05dfdf18-b064-11df-8c04-00144feabdc0.html#axzz1EUGI2ouk
Muthu, S and R Raman (2006) Concentrated Paraboloid Solar Cookers for Quantity Cookery, Paper presented at *Solar Cookers International Conference*, 12–16 July 2006, Grenada, http://solarcooking.org/Granada06/43_satyavathi_muthu.pdf

Narain, U and A Krupnick (2007) The Impact of Delhi's CNG Program on Air Quality, *Resources for the Future Discussion Paper*. RFF DP 07-06.

National Environment Agency (2010a) *Singapore's Second National Communication under the United Nations Framework Convention on Climate Change*, Singapore: National Environment Agency of Singapore, http://app.nccs.gov.sg/data/resources/docs/SINGAPORE'S%20SECOND%20NATIONAL%20COMMUNICATIONS%20NOV%202010.pdf

National Environment Agency (2010b) *Solid Waste Management in Singapore*, 2nd Regional 3R Forum in Asia, Singapore: Singapore Government.

National Environment Agency (2012) *National Environment Agency*, Singapore: Singapore Government, http://www.nea.gov.sg/

National Parks (2009) Bedok Reservoir Park, *National Parks*, Singapore: National Parks Board, http://www.nparks.gov.sg/cms/index.php?option=com_visitors guide&task=parks&id=6&Itemid=73.

National Parks Board (2012a) *Park Connector Network*, NParks, Singapore: Singapore Government, http://www.nparks.gov.sg/cms/index.php?option=com_visitorsguide&task=parkconnectors&Itemid=74

National Parks Board (2012b) *Skyrise Greenery Award*, NParks, Singapore: Singapore Government, http://www.skyrisegreenery.com/index.php

National University of Singapore (2009–2010) *Report Paper on Singapore — A Potential Carbon-Neutral City, What Will Be the Next Design and Development Phase for Singapore in the Age of Sustainability?* Singapore: Master of Arts (Urban Design), National University of Singapore.

Newman, P (2006) The Environmental Impact of Cities, *Environment and Urbanization* 18(2): 275–295.

Newman, P (2009) Planning and Sustainable Urban Development: Linking the Green and Brown Agendas, Chapter 6 in UN Habitat (2009) *Planning Sustainable Cities*, Global Report on Human Settlements, United Nations Habitat, Nairobi.

Newman, P (2011) Green Urbanism and its Application to Singapore, *Asia Research Institute Working Paper* Series 151, March 2011, National University of Singapore.

Newman, P (2012, 14 May) King Coal Dethroned, *The Conversation*, http://theconversation.edu.au/king-coal-dethroned-6977

Newman, P and A Matan (2012) Human Health and Human Mobility, *Current Opinion in Environmental Sustainability* 4(4): 420–426, http://dx.doi.org/10.1016/j.cosust.2012.08.005

Newman, P and I Jennings (2008) *Cities as Sustainable Ecosystems: Principle and Practices*, Washington DC: Island Press.

Newman, P and J Kenworthy (1999) *Sustainability and Cities: Overcoming Automobile Dependence*, Washington, DC: Island Press.

Newman, P and J Kenworthy (2007) Greening Urban Transport, *State of the World, 2007*, Washington, DC: World Watch Institute, pp. 68–85.

Newman, P and J Kenworthy (2011) 'Peak Car Use': Understanding the Demise of Automobile Dependence, *World Transport Policy and Practice* 17(2): 31–42.

Newman, P, T Beatley and H Boyer (2009) *Resilient Cities: Responding to Peak Oil and Climate Change*, Washington, DC: Island Press.

Newman, P, T Beatley and L Blagg (2012) *Singapore: Biophilic City* [Film], Curtin University Sustainability Policy (CUSP) Institute and Sustainable Built Environment National Research Centre (SBE), http://www.youtube.com/watch?v=XMWOu9xIM_k

Nielsen (2012) *The Global, Socially-Conscious Consumer*, March 2012, The Nielsen Company, http://www.nielsen.com

Ninbo University (2008) *Ninbo University*, http://oldweb.cqvip.com/qk/85154X/201006/34212203.html [non-English language website].

Nolan, RB and LL Lem (2001) *A Review of the Evidence for Induced Travel and Changes in Transportation and Environmental Policy in the United States and the United Kingdom*, London: Centre for Transport Studies, Imperial College.

Norton, M (2007) *Japan's Eco-towns — Industrial Clusters or Local Innovation Systems?* Proceedings of the 51st Annual Meeting of the ISSS Tokyo Institute of Technology, Tokyo, Japan, August 5–10 2007.

Organization for Economic Co-operation and Development (OECD) (2011a) *Better Policies for Better Lives*, OECD Secretary-General Angel Gurría, OECD Week, Paris: OECD.

Organization for Economic Co-operation and Development (OECD) (2011b) *Towards Green Growth*, Paris: OECD.

Organization for Economic Co-operation and Development (OECD) (2012) *Compact City Policies: A Comparative Assessment*, OECD Green Growth Studies, OECD Publishing.

Oh, I, W Wehrmeyer W and Y Mulugetta (2010) Decomposition Analysis and Mitigation Strategies of CO_2Emissions from Energy Consumption in South Korea, *Energy Policy* 38(1): 364–377.

Oppenheimer, M (2010, 20 March) Masdar — Future Renewable Energy City, *iCentrus*, Renewable energy, http://www.icentrus.com/masdar-future-renewable-energy-city/

Osanai, T (2009) The Green School, *Outdoor Japan*, from http://outdoorjapan.com/magazine/story_details/69

Pandey, M (2012) Choked: Delhi's Pollution Level is Five Times Worse than Just Eight Years Ago as Experts Warn Air Quality is as Noxious as Jharia Coal Town, *Daily Mail*, http://www.dailymail.co.uk/indiahome/indianews/article-2170277/Choked-Delhis-pollution-level-fives-times-worse-just-years-ago-experts-warn-air-quality-noxious-Jharia-coal-town.html

Park Kil-Dong (n.d.) *Cheonggyecheon Restoration Project*, Korea: Seoul Metropolitan Government, http://www.wfeo.org/

Pasig River Rehabilitation Commission (2012) *Pasig River Rehabilitation Commission*, http://www.prrc.gov.ph/

Philippine Green Movement (2009, 13 March) Philippines Environmental Issues: How did the Pasig River get polluted? *Philippines Green Movement*: Environmentalism in the Philippines, http://www.thegreentheory.com/philippines-environmental-issues-how-did-the-pasig-river-get-polluted

Pohl, EB (2009, 31 August) Masdar Sustainable City/LAVA, *ArchDaily*, http://www.archdaily.com/33587/masdar-sustainable-city-lava/

Preston, B and R Jones (2006) *Climate Change Impacts on Australia and the Benefits of Early Action to Reduce Global Greenhouse Gas Emissions*, A Consultancy Report for the Australian Business Roundtable on Climate Change, Canberra, Australian Capital Territory: CSIRO.

PRLog (2010, 29 December) Xianyang Geothermal Field — Shaanxi, *PRLog*, http://www.prlog.org/11182732-xianyang-geothermal-field-shaanxi.html

Proconservation (2010) Kampung Kali Chode, We Conserve, You Conserve, http://proconservation.blogspot.com.au/2010/01/kampung-kali-chode-yogyakarta-indonesia.html

Property Wire (2009, 27 April) Carbon Neutral Development a First for Central Asia, *Property Wire*, http://www.propertywire.com/news/asia/carbon-neutral-development-asia-200904273009.html

Public Utilities Board (2010) *Public Utilities Board*, Singapore: Singapore Government, http://www.pub.gov.sg/

Public Utilities Board (2012) *NEWater*, Singapore: Singapore Government, http://www.pub.gov.sg/water/newater/Pages/default.aspx

Puerto Princesa City (2010) *Puerto Princesa City*, http://puertoprincesacity.net/index.html

Puig, J (2005) Energy Efficient Cities: Political Will, Capacity Building and Peoples' Participation. The Barcelona Solar Ordinance: A Case Study about How the Impossible Became Reality. In P Droege (Ed) *Urban Energy Transition*, New York: Elsevier.

Purohit, P and A Michaelowa (2007) Potential of Wind Power Projects under the Clean Development Mechanism in India, *Carbon Balance and Management* 2(8), http://www.cbmjournal.com/content/2/1/8

Putnam, R (1993) *Making Democracy Work: Civic Traditions of Modern Italy*, Princeton, NJ: Princeton Architectural Press.

Rahman, A (1999) *Women and Micro-credit in Bangladesh*, United States of America: West View Press.

Renewable Power News (2009, 5 November) Taiwan Proud Owner of Largest Solar Powered Stadium, *Renewable Power News*, http://www.renewablepowernews.com/archives/291

Republic of China (Taiwan) (2010) *Renewable Energy*, Republic of China (Taiwan), Government Information Office, Republic of China (Taiwan), http://www.taiwan.gov.tw/mp.asp?mp=999

Reuters and D Gray (2010) A Monument, Which Reads: "Home of Rare Earths Welcomes You", Stands in a Field of Wind Turbines Near the Town Of Damao in China's Inner Mongolia Autonomous Region, October 31, 2010, *Reuters*, http://www.reuters.com/

Revkin, A (2008, 5 February) Car-Free, Solar City in Gulf Could Set a New Standard for Green Design, *The New York Times*.

RIBA Journal (2009) China Blueprint, *RIBA Journal*, http://www.ribajournal.com/index.php/feature/article/china_blueprint_AUGSEPT09/

Roberts, B and T Kanaley (Eds) (2006) Urbanization and Sustainability in Asia: Case Studies of Good Practice, *Asian Development Bank*, http://www.adb.org/Documents/Books/Urbanization-Sustainability/urbanization-sustainability.pdf

Rosencranz, A and M Jackson (2002) Clean Air Initiative for Asian Cities, The Delhi Pollution Case, New Delhi, http://www.indlaw.com/

Rubin, J (2009) *Why Your World is About to Get a Whole Lot Smaller: Oil and Globalisation*, New York: Random House.

Saieh, N (2009, 8 June) Tulou Housing Guangzhou/URBANUS Architects by Iwan Baan, *Archdaily*, http://www.archdaily.com/24210/tulou-housing-guangzhou-urbanus-architects-by-iwan-baan/

Saieh, N (2010, 13 October) The Green School/PT Bambu, *ArchDaily*, http://www.archdaily.com/81585/the-green-school-pt-bambu/

Sanyo (2012) Solar Ark Gallery, *Sanyo Electric Co.*, Ltd., Panasonic, http://sanyo.com/solarark/en/about/

Sawin, JL and K Hughes (2007) Energizing Cities, *State of the World, 2007*, Washington, DC: Worldwatch Institute, pp. 90–107.

Scheurer, J (2003) *Urban Ecology, Innovations in Housing Policy and the Future of Cities: Towards Sustainability in Neighbourhood Communities* (PhD Thesis), Institute for Sustainability and Technology Policy, Perth: Murdoch University.

Scheurer, J and P Newman (2008) *Vauban: A Case Study in Public Community Partnerships*, Case Study for United Nations Global Review of Human Settlements, http://www.sustainability.curtin.edu.au

Schuiten, L (n.d.) *Luc Schuiten — Atelier d'Archictecture Schuiten*, Vegetal City, http://vegetalcity.net/

Seagate (2008) Sakai City Waterfront Mega Solar Power Generation Plan, *Flicker*, http://www.flickr.com/photos/seagate/2620760470/in/photostream/

Seoul Metropolitan Government (2011) Destinations: Cheonggyecheon (Stream), *Seoul Metropolitan Government*, http://english.seoul.go.kr/cav/att/chenggye.php

South East Queensland Healthy Waterways Partnership (2010) 'What is Water Sensitive Urban Design?', *Water By Design*, http://waterbydesign.com.au/whatiswsud/

Setiawati, I (2010, 3 June) City to Start Building Sewage Tunnel System in 2011, *The Jakarta Post*, http://www.thejakartapost.com/news/2010/03/06/city-start-building-sewage-tunnel-system-2011.html

Shanghai Shentong Metro Group Co., Ltd (2010) Shanghai Metro, http://www.shmetro.com/

Sharp (2008, 23 June) Press Release, Sakai City, *Sharp*, http://sharp-world.com/corporate/news/080623_2.html/

Singapore Government (n.d.) *From Garden City to City in a Garden*, Singapore: Singapore Government.

Singh, T (2011, 29 March) Masdar City: A Carbon-neutral Metropolis, *MENA Infrastructure*, http://www.menainfra.com/news/masdar-city-carbon-neutral-/

Sino-Singapore Tianjin Eco-City (n.d.) *Sino-Singapore Tianjin Eco-City*, Singapore: Singapore Government, http://www.tianjinecocity.gov.sg

Sirolli, E (1999) *Ripples from the Zambezi: Passion, Entrepreneurship, and the Rebirth of Local Economies*, Gabriola Island, BC: New Society Publishers. See also http://www.sirolli.com/

Six Battery Road (2011) *Six Battery Road*, Capita Commercial Trust Management, http://www.sixbatteryroad.com/

Sobo, D and Z Hoberg (2010) Report Paper on Sustainable Living in Auroville, India, *Auroville*, http://www.auroville.org/research/AV_Sustainability_Study_Project_by_Visiting_Students.pdf

Solar Energy Demystified (2010, 8 March) Solar Steam Cooking System at Tirumala, *Solar Energy Demystified*, http://solarenergydemystified.wordpress.com/2010/03/08/solar-steam-cooking-system-at-tirumala/

Solar Power — PV Panels (2012) Japan's Solar Powered City, *Solar Power — PV Panels*, http://solarpowerpanels.ws/solar-power/japans-solar-powered-city

Songdo IBD (2012) *Song do IBD*, Gale International, http://www.songdo.com/songdo-international-business-district/why-songdo

South Fremantle Senior High School (2012) *South Fremantle Senior High School*, http://www.skyline-creations.com/schools/sfshs/html/our_school.html

Southeast Asia Building (2011) An Unexpected 'Hanging Garden' in the CBD, *Southeast Asia Building* September/October (2011): 51–55.

Standing Advisory Committee on Trunk Road Assessment (1994) *Trunk Roads and the Generation of Traffic*, London: Department of Transport.

Stanton, C (2011, 11 October) Masdar City Overhaul Cuts $3.3 Billion from Costs, *The National*, http://www.thenational.ae/business/energy/masdar-city-overhaul-cuts-3-3bn-from-costs

Starrs, T (2005) The SUV in Our Pantry, *Solar Today*, http://www.sustainablebusiness.com/index.cfm/go/news.feature/id/1275

Stein A (2009, 17 July) Urban Waterways: Seoul Peels back the Pavement and Reveals a River, *The TerraPass Footprint*, http://www.terrapass.com/blog/posts/seouls-river

Suniva Inc. (2010) Rooftop of India's First Eco-Friendly Stadium, 1MW+, New Delhi, Powered by Suniva, *Flicker*, http://www.flickr.com/photos/suniva_solar/4798588467/

Sustainable Cities (2011) Seoul: Life beneath the Asphalt, *Sustainable Cities*, Danish Architecture Center, http://sustainablecities.dk/en/city-projects/cases/seoul-life-beneath-the-asphalt.

Sustainable Cities (n.d.) Rizhao: Mainstreaming Solar Energy on City Level, *Sustainable Cities*, Danish Architecture Centre, http://sustainablecities.dk/en/city-projects/cases/rizhao-mainstreaming-solar-energy-on-city-level

SWITCH Asia Programme (2011) *Building Energy Autonomous Resorts Creating Appropriate Technology Solutions*, UNEP/Wuppertal Institute Collaborating Centre on Sustainable Consumption and Production, http://www.switch-asia.eu/switch-projects/project-progress/projects-on-improving-production/zero-carbon-resorts.html

Taiwan Sustainable Cities (2011) Kaohsiung: Taiwan's Eco-City Leader? *Taiwan Sustainable Cities*, http://taiwansustainablecities.blogspot.com/2011/01/kaohsiung-taiwans-eco-city-leader.html/

Tan, R (2012a) *Wild Shores*, http://wildshores.blogspot.com.au/

Tan, R (2012b) *Wild Singapore*, http://www.wildsingapore.com/index.html

Tay, E (2010, 20 March) Singapore to Become a Smart Energy Economy, *Lowcarbonsg*, Green Future Solutions, http://www.lowcarbonsg.com/category/issues-and-policies/

Tay, E (2011, 19 January) Singapore's Second National Communication on Climate Change Report to the UNFCCC Secretariat, *Lowcarbonsg*, Green Future Solutions, http://www.lowcarbonsg.com/2011/01/19/singapores-second-national-communication-on-climate-change-report-to-the-unfccc-secretariat/

TCRP (2003) *Bus Rapid Transit — Volume 2: Implementation Guidelines*, Report 90, TCRP.

The Aga Khan Award for Architecture (n.d.) *Grameen Bank Housing Programme*, http://www.akdn.org/architecture/pdf/1066_Ban.pdf

The Star (2009, 24 January) *The Star*, http://biz.thestar.com.my/archives/2009/1/24/business/bw_p26ZEO.jpg

Thomas, J (2008, 10 May) Korean Village Runs on 100% Solar Power, *Metaefficient*, http://www.metaefficient.com/renewable-power/korean-village-runs-on-100-solar-power.html

Trekthailand (n.d.) National Parks in Myanmar (Burma), Hlawaga park — Yangon, *Trekthailand*, http://www.trekthailand.net/myanmar/parks/hlawaga/

United Nations and Asian Development Bank (2012) *Green Growth, Resources and Resilience: Environmental Sustainability in Asia and the Pacific*, ST/ESCAP/2600, RPT124260, Bangkok: United Nations and Asian Development Bank Publication.

United Nations, Bureau International des Exhibitions and Municipal Government of Shanghai (2011) *Shanghai Manual: A guide for Sustainable Urban Development of the 21st Century. Better City, Better Life*, United Nations, Bureau International des Exhibitions and Municipal Government of Shanghai, http://www.un.org/esa/dsd/susdevtopics/sdt_humasett_capacitybuildings.html

United Nations Development Programme (UNDP) (2010) *Country Summary Report: Bangladesh Project for Increasing Stakeholder Utilization of GAR 11 Preparation Study for Asia*, Practical Action for UNISDR February 2010, United Nations Development Programme, http://www.undp.org.bd/info/pub.php

United Nations Environment Programme (UNEP) (2010) *Overview of The Republic of Korea's National Strategy for Green Growth*, April 2010, United Nations: United Nations Environment Programme, http://www.unep.org/PDF/Press Releases/201004_unep_national_strategy.pdf

United Nations Environment Programme (UNEP) (2012) *Convention on Biological Diversity*, United Nations Environment Programme, http://www.cbd.int/

United Nations Environment Programme and Bloomberg New Energy Finance (2011) *Global Trends in Renewable Energy Investment 2011: Analysis of Trends and Issues in the Financing of Renewable Energy*, http://www.newenergyfinance.com/

United Nations Framework Convention on Climate Change (UNFCCC) (2006) *Clean Development Mechanism Project Design Form*, Version 3, Clean Development Mechanism, http://cdm.unfccc.int/

United Nations Framework Convention on Climate Change (UNFCCC) (2007) Surat Thani Biomass Power Generation Project in Thailand, *Clean Development Mechanism Project Design Form*, Version 3, Clean Development Mechanism.

United Nations Human Settlements Programme (UN-Habitat) (2009) *United Nations Review of Human Settlements*, Chapter 6 Integrating the Green and Brown Agendas, Nairobi, Kenya: UN Habitat.

United Nations Human Settlements Programme (UN-Habitat) (2012) Urban Patterns for a Green Economy: Working with Nature, http://www.unhabitat.org.

Urban Transportation Monitor (1999) Summary Information from Texas Transportation Institute Annual Mobility Report, *Urban Transportation Monitor* 13(22), Burke, Virginia: Lawley Publications.

Urbane Community (2010) *One Village One Playground Report*, Urbane Community.

Verkís (n.d.) Xianyang Geothermal District Heating — China, *Verkís*, http://www.verkis.com/projects/energy/geothermal/nr/1420

Vidal, J (2006, 1 November) Heart and Soul of the City, *The Guardian*, http://www.guardian.co.uk/environment/2006/nov/01/society.travelsenvironmentalimpact

Virtualtourist (n.d.) *Virtual tourist*, http://www.virtualtourist.com/

Wang, EE (2009, 26 October) Singapore's First "Zero Energy Building" Launched, *Wild Singapore*, http://wildsingaporenews.blogspot.com/2009/10/singapores-first-zero-energy-building.html

Waste Management and Recycling Association of Singapore (2001) *Waste Management and Recycling Association of Singapore, Market Survey*, Singapore: Waste Management and Recycling Association of Singapore.

Waste Management World (2011) Semakau Landfill, *Waste Management World*, http://www.waste-management-world.com/index/display/article-display/356697/articles/waste-management-world/landfill/2009/03/semakau-landfill.html

Waypoints (2007) *Bangui Windmills*, http://www.waypoints.ph/detail_gen.php?wpt=windf1

Webster, A (2012, 17 January) Chinese Aims to Cut Carbon Intensity by 45%, Good Clean Tech, http://goodcleantech.pcmag.com/none/292932-chinese-aims-to-cut-carbon-intensity-by-45

Went, A, W James and P Newman (2008) *Renewable Transport: Integrating Electric Vehicles, Smart Grids and Renewable Energy*, CUSP Discussion Paper, Curtin University Sustainability Policy Institute, Fremantle, http://sustainability.curtin.edu.au/local/docs/cusp_discussion_paper.pdf

Williams, J (2012) *Zero Carbon Homes: A Road Map*, Oxon: Earthscan.

Wilson, EO (1984) *Biophilia*, Cambridge, MA: Harvard University Press.

World Architecture Community (2012) *World Architecture Community*, http://www.worldarchitecture.org

World Bank (2008, August) *The World Bank in Thailand Newsletter*, World Bank.

World Bank (2012) *The World Bank*, The World Bank Group, http://www.worldbank.org/

World Future Council (n.d.) Sustainable Cities Case Study Rizhao, *Area: Africa Renewable Energy Alliance*, http://www.area-net.org/fileadmin/user_upload/AREA/AREA_downloads/Policies_SolarWaterHeater/WFC_Case-Study_China-Rizhao.pdf/

World Health Organisation (2011) Indoor Air Pollution and Health, Fact Sheet No 292, September 2011, http://www.who.int/mediacentre/factsheets/fs292/en/

World of Renewables (2010, 17 February) Puerto Princesa, PhilBio Groundbreak Waste-to-energy Project, *World of Renewables*, http://www.worldofrenewables.com/renewables_news/wastetoenergy/puerto_princesa_philbio_groundbreak_waste-to-energy_project.html

World Wind Energy Association (2006) Wind Farms around the World, *World Wind Energy Association*, http://www.wwindea.org/home/index.php?option=com_content&task=view&id=83&Itemid=2

Xi, F, Y Geng, X Chen, Y Zhang, X Wang, B Xue, H Dong, Z Liu, W Ren, T Fujita and Q Zhu (2011) Contributing to Local Policy Making on GHG Emission Reduction through Inventorying and Attribution: A Case Study of Shenyang, China, *Energy Policy* 39: 5999–6010.

Xin, D (2008, 17 November) Blueprint of Railways Development, *China Daily*.

Xu, X (2010, 18 January) From Solar Dreams to an Energetic Reality, *China Daily*, http://www.chinadaily.com.cn/cndy/2011-01/18/content_11870429.htm

Yanlord Land Group, Sembcorp Industries and Sembcorp Parks Management (2009, 25 May) *Singapore Consortium to Lead Development of Nanjing City's Largest Foreign Collaborative Project 'Sino-Singapore Nanjing Eco High-Tech Island'*, Press Release, Yanlord Land Group Limited, Sembcorp Industries Ltd and Sembcorp Parks Management Pte Ltd, http://www.finanznachrichten.de/pdf/20090525_124549_Z25_20ABBBFEDDE03CAB482575C10015426C.2.pdf

Yok, TP and A Sia (2008) *A Selection of Plants for Green Roofs in Singapore* (2nd Ed), Singapore: Centre for Urban Greenery and Ecology, National Parks Board.

Yok, TP, B Yeo, YW Xi and LH Seong (2009) *Carbon Storage and Sequestration by Urban Trees in Singapore*, Singapore: Centre for Urban Greenery and Ecology, National Parks Board.

Yoneda, Y (2011, 15 August) The World's Largest Solar Energy Office Building Shines in China, *Inhabitat*, http://inhabitat.com/worlds-largest-solar-energy-office-building-opens-in-china/

Young, S (2012) Building the Architecture for Green Growth, *Making It: Industry for Development*, 9, United Nations Industrial Development Organization (UNIDO),http://www.makingitmagazine.net/?p=4510

Zudianto, H and D Parikesit (n.d.) *Promoting Livable City: Integrating local/global environmental concerns and local development challenges*, http://www.pustral-ugm.org

About the Authors

Professor Peter Newman is John Curtin Distinguished Professor of Sustainability and Director of Curtin University Sustainability Policy (CUSP) Institute in Australia. He is on the advisory council of Infrastructure Australia, is a Lead Author for the IPCC and is the author of 15 books on sustainability and cities.

Dr Anne (Annie) Matan is a researcher and lecturer at Curtin University Sustainability Policy (CUSP) Institute in Australia, interested in creating sustainable, vibrant and people-focused urban places. Her research focus is on walkability, pedestrian planning, urban design and transportation planning.

Index